MW00575744

Praise for *In Search of the Perfect Peach*

'Franco and Natoora's pioneering approach to flavour-led food has changed food for the better in the UK and beyond. This book delves further into what eating for flavour means for our plates and the planet.'

Anna Jones, cook and bestselling author of *Easy Wins*

'The current food system is causing unique flavours to be lost and our palates distorted. *In Search of the Perfect Peach* gives a powerful and impassioned account of why this is harmful to people and planet. By valuing and reclaiming flavour, Franco argues that we *can* transform the system and also enrich our relationship with food. Essential reading.'

Dan Saladino, journalist,
broadcaster and author of *Eating to Extinction*

'The growth of The River Cafe in London over the past twenty years has been made possible by Natoora, who bring our ingredients in the right season, from the right region, with the best quality and taste to our door. This exciting and important book now brings Natoora and all its ingredients not just to restaurants like ours but also to the interested cook, to their home – to everyone's home.'

Ruth Rogers, chef and co-founder of The River Cafe

'An incredible read for those who love food and care about its future. *In Search of the Perfect Peach* is both a love story and technical guide to supporting good food systems and the people behind them.'

Kyle Connaughton, chef and owner of
SingleThread Farm, Restaurant and Inn

In Search of the Perfect Peach

In Search of the Perfect Peach

Why flavour holds the answer to fixing our food system

Franco Fubini

Foreword by Tim Spector

Chelsea Green Publishing
London, UK
White River Junction, Vermont, USA

Commissioning Editor: Muna Reyal
Project Manager: Susan Pegg
Copy Editor: Susan Pegg
Proofreader: Sarah Greaney
Indexer: Charmian Parkin
Page Layout: Laura Jones-Rivera

Printed in the United States of America.
First printing July 2024.
10 9 8 7 6 5 4 3 2 1   24 25 26 27 28

ISBN 978-1-915294-29-6 (hardback)
ISBN 978-1-915294-30-2 (ebook)
ISBN 978-1-915294-31-9 (audiobook)

Library of Congress Control Number: 2024941510

Chelsea Green Publishing
London, UK
White River Junction, Vermont, USA
www.chelseagreen.co.uk

To my mother, Eva.

For leading me on the path to flavour and a life full of meaning.

Contents

Foreword

Diversity is vital for life. This simple phrase lies at the heart of what both Natoora and I do. We approach our work from different directions, but our conclusions are identical.

Our planet is host to an incredibly complex set of ecosystems that are known as biomes. Life has evolved amidst vast diversity on both the macro and micro scales, from the innumerable microorganisms that first ignited evolution to the incredible abundance of organisms and ecosystems that comprise Earth.

All animals originally evolved from simple microorganisms. These single-celled creatures also evolved alongside us and depended on us for food and shelter. But today, these relationships that have underpinned life on Earth across aeons are being ripped apart.

The intricate webs that bind species together are slowly being degraded. If these webs fail (a real possibility if we continue on our current course), the damage to our health and the health of the planet will be irreparable.

Modern agriculture is stripping our soil of nutrients and microbes; the food industry is stripping our food of its nutrient density; and the Western diet is stripping our own ecosystem, our gut microbiome, of its diversity. These losses of diversity mean that our combined future is on the line.

Much of my research now focuses on the gut microbiome – the trillions of microorganisms that live in our gut. Just a few decades ago, we thought these microbes helped us to digest food, gave us food poisoning and did little else of interest.

Now, we know so much more. The links between the gut microbiome and almost any health condition you can think of are clear. This is because gut bacteria heavily influence our immune system by 'training' it, regulating inflammation and allergies, reducing ageing and cancer, and ensuring we can thrive. They even play key roles in our mood and mental health.

A key measure of the health status of the gut microbiome is diversity. The Hadza people in Tanzania have some of the most diverse gut microbiomes on the planet and the fewest diseases. I was lucky to spend some time with these wonderful people. After just three days following their hunter-gatherer lifestyle, my gut microbial diversity increased by an incredible 20 per cent.

The foods I ate and the way they were grown no doubt had a huge influence, but the act of foraging itself and being exposed to microbes in the environment was equally important. In the Western world, there is far too much focus on sterility: we have created environments that are rich in plastics, pesticides and other chemicals that we expose ourselves to through our skin and lungs, especially in cities. No longer do we live in harmony with the natural world and all its wonderfully diverse microbial forms of life. Only around 1 per cent of bacteria cause disease, while the rest are safe or actively beneficial.

Sadly, just a few days after my Tanzanian experience, my microbiome returned to baseline. This demonstrates both the incredible flexibility of the gut microbiome and its sensitivity to environmental and dietary changes. In the Western world, diversity is declining. As there is on a macro level on our planet, there is a mass extinction event going on in our guts, with an estimated 50 per cent of microbe species now permanently lost.

What can we do about this? Seven years ago, I co-founded ZOE, a science and nutrition company. The programme includes a gut microbiome test and we guide our members through healthy, daily dietary choices suited to their metabolism. We now offer a repeat gut microbiome test after a few months. This data shows without a doubt that a diverse, plant-based diet with limited ultra-processed foods (UPFs) improves the gut microbiome. We see a reduction in 'bad' species related to poor health outcomes like obesity, heart disease and inflammation and a marked increase in 'good' species associated with improved health.

Faced with the meteoric rise in diet-related diseases like cancer and type 2 diabetes, we aim to improve the health of millions. However, we have a battle on our hands. Big Food holds all the cards. The sharp rise in UPF (fake food) consumption is deeply worrying. In the United States and United Kingdom, 60 per cent of adults' calories are consumed via these edible food-like substances and over 70 per cent of children's. Fake foods rarely contain fibre, which our gut bacteria rely on. They do contain sugar, unhealthy fats and a vast array of chemicals, like sweeteners and emulsifiers.

Additives, colourants, artificial flavourings and the like have all been tested for safety. But many of these tests were conducted decades ago on rodents and checked individual chemicals, not the cocktails found in our food. They also mostly focused on cancer.

Studies looking at the impact of these chemicals on gut health are rare. And the food industry, of course, has no interest in investigating their additives any further than the law requires, nor are they obligated to do so. The reality is that no one knows what happens when you consume complex cocktails of these chemicals every day for a lifetime. Multiple epidemiology studies now show that eating fake foods in the long term increases the risk of nearly every common disease, especially cancers and heart disease, by up to 50 per cent. Much of this effect can come from damage to the gut microbes, either from starving them of fibre or as a direct result of the chemicals. Recent research has confirmed that both emulsifiers and sweeteners harm the gut microbiome.

Recently, I urged the UK government to fight back against the food giants who are, in a very real way, poisoning populations and making obscene profits. Big Food, with its vastly deep pockets, employs lobbyists to slow down change and massage food laws for their benefit. The fact that they spend millions, or even billions, on lobbying each year means that it must work. The system is rigged in their favour.

We now know that eating a diverse range of plants – preferably at least thirty types each week, which can include fruits, vegetables, nuts, seeds, herbs or spices – is linked to improved markers of long-term health and a thriving, diverse gut microbiome. Yet the global food industry, despite its vast selection of fake food products, is preventing us from reaching this goal.

Today, just three plants – wheat, rice and corn — account for 60 per cent of our calories. While the food industry makes a killing, farmers are increasingly squeezed to ensure that supermarkets and food manufacturers can maximise profits with these monocultures. Farming has become agribusiness, focused on extracting profit rather than stewarding the land it uses, the environment it exists in and caring for the people it feeds.

We need to take stock of the current situation and, as individuals and active participants within the food system, call for and demonstrate what we need to galvanise the necessary change. Our daily food choices are the strongest tool we have, like our right to vote, to make our voices heard.

Collectively as consumers we yield great power. As Franco reveals in this book, what we decide to eat and the demand we foster has a direct impact on the development of what food is grown and how it moves from farm to plate. We can influence, one producer at a time, one farm at a time, those individuals who look after our soils and protect the valuable ecosystems. By choosing diversity in our diets, we will in turn demand diversity from our food system – this is good for the planet's health, but also the health of our own microbiome. Following flavour is a source of joy that leads us on a path towards a more natural relationship with our food. It signposts that what you are eating has been grown on healthy soil rather than one artificially enriched by chemical fertilisers. In turn, the food we grow will be more nutritious. Everyone will benefit.

Approaching diversity from the perspective of the craft of growing and the joy of eating, Franco and Natoora are building a new path. Not just to protect the diversity of life on Earth, but to protect the diversity of life within us. Let's all raise a glass to diversity!

Tim Spector
Professor of genetics,
author and scientific co-founder of ZOE
July 2024

INTRODUCTION

The Story of a Peach

In the summer of 2011, I bit into a Greta white peach for the first time. It was early morning and a beautiful dawn light was casting itself across the halls of the market in Milan, a light that exposed the raw beauty of the Greta's skin.

I could see it was a white peach – the flesh's colouration always comes through to the surface – but the skin on this peach was unlike anything I had ever seen before. It was an artist's monochromatic study in shades of purple, crimson-fire blood on the inside, with the most astonishing character. Biting into it was like taking off on a rocket ship – its acidity edged just over its sweetness into a flavour so intense and magical that it blew my mind and seared itself forever in my memory. The whole experience was otherworldly. Biting into a great ripe peach... Well, that is pure joy.

It sparked an obsession with flavour that led me to search for the grower of the Greta. I wanted to meet this artist, to understand how he was capable of coaxing nature, working with her, to produce these magnificent works of edible art. I had a thirst for knowledge: I wanted to, needed to understand the circumstances that led to each crimson brushstroke, and eventually, though it took me a few years, I did find the farmer behind these works of art. Years later, I would discover a strong connection between flavour and nutrition, and I realised that my fixation on flavour had far-reaching positive consequences for our health. There is also a strong chance that food that is full of flavour has been farmed in a way that heals our planet.

Since founding Natoora, I've spent the last twenty years sourcing the finest fruits and vegetables for home cooks as well as some of the world's best chefs. Although this is just one story of many, the journey it took me on has become emblematic of how flavour has spurred me into a lifelong quest to fix the food system: one bite led me on a wild goose chase to seek out a peach variety with a two-week season that eventually turned into a ten-year sourcing relationship with Domenico, a third-generation stone fruit grower in Campania.

What is important about this is that flavour is not just a momentary, visceral, personal experience but a window into the farmer, the soil and the entire ecosystem. A window that will help us to become better-informed cooks and consumers.

This book is the culmination of two decades of hard graft and deep thinking. I have dedicated a lot of my time to considering both our food system and what is required to fix it, along with the practical experience of building Natoora – a wholesaler, retailer and producer of seasonal food. Obsessively focused on flavour, at Natoora we work with hundreds of incredible farms around the world with a mission to revolutionise the food system. In the pages that follow, I make the case for flavour and how by using it as a tool to guide our food choices we can achieve this revolution.

I will explain how real flavour is created and its importance beyond the joy it elicits, as seen through the lens of nutrition and soil health. I discuss how to recognise flavour, allowing us to find good food and appreciate it. The stories I have chosen to bring together in this book draw on the relationships I've built with farmers and producers across the world over the years: to honour flavour is to first honour their craft. My hope is that these stories will serve as a source of inspiration to bring back a much-needed reverence for flavour and the commitment required to produce it. But more importantly, their interconnectedness links all the arguments in the book back to what matters most: human and planetary health. And in doing so, allowing us to understand that flavour is the key to understanding good food and understanding our food system, and that gives us the power to revolutionise it.

Culture and Food Memories

It is no secret today that the industrial food system, which governs the vast majority of the food we consume, is an aberration of significant proportions, overfeeding those who have too much, leaving millions undernourished, and destroying both our soils and our natural environment. My search for flavour didn't start there, however.

As humans evolved over tens of thousands of years, so did our culture and alongside it our connection to food and the land. Food moved from pure sustenance to community and enjoyment. We celebrate with it, and around it we build the human connections that hold families and communities together. The act of cooking for others is one of the greatest pleasures in life when done with love, thought, creativity and great produce. Flavour was instrumental in this evolution. If food were simply fuel with no joy, no pleasure, it would not hold such prominence in our traditions and celebrations. Flavour was the spark and without it food would be but a footnote in our culture.

It is with memories, my own sense of cultural history, that my personal search for flavour began. Memories, particularly those from childhood, are immensely formative. Our enjoyment of food is bound to those early experiences. Your idea of the perfect tiramisu is informed by what you had as a child: the cream must be runny, not set; the ladyfingers not overly soaked. Yet to someone else, it might be quite the opposite. Across the world, across cultures and families, this is how food memories work. Our idea of the perfect dish is all down to what we had as a child. Our love of burnt rice on the bottom of a pan, or what the right amount of spice is. We can't erase those early experiences.

I was always interested in food, always drawn to produce. My parents separated early on in my childhood when we were living in Argentina, and a few years later my mother, my brother and I moved to Italy. It was there she met my stepfather, Alberto, who at the time was the consul general at the Argentinian consulate in Milan. Alberto was an incredible man and I learned as much from seeing the love he had for my mother as I did from his unwavering conduct. Although he sadly passed away too soon, the short years we lived with him were intense in the best of ways and made me a far better person. He imprinted in my brother and me a deep sense of the value of family, moral fortitude, honesty and integrity – values that have been our core at Natoora. He was also responsible for moving our family around the world as we followed his diplomatic career, giving me an invaluable experience and exposing me from an early age to various different cultures. I vividly remember the insane mangoes we had in Egypt and the recipe my mother picked up for mango chutney. Fortunately, the countries I grew up in had a strong food culture, places like Italy, the Netherlands, Argentina and Egypt. Industrialisation back in the late 1970s and 80s had not reached these countries – supermarket culture was in its infancy – so I had access to varieties of produce, fruits and vegetables that were bred for flavour and tasted amazing.

Our summers in southern Italy were full of phenomenal tomatoes. My brother, Ale, and I used to go to this homey deli steps from the beach in Positano and get Parma ham, mozzarella and tomato sandwiches. That memory is why, to this day, this is one of the best sandwiches for me – it is pure perfection with a little salt and olive oil.

In Holland I discovered Gouda cheese, which my brother and I used to slice thinly and eat with crackers after school. I even remember a small, very pale green fruit that looks like a mini nectarine, which I found at a supermarket in the town of Maldonado in Uruguay on a summer holiday. Thirty-five years later and I still remember where I took that first bite, how it looked and tasted. I've never found that fruit again. As I grew older and left home, I carried that life with me, those experiences and wonderful, joyous memories.

I've never been far from the kitchen. At home, my mother always cooked, every single night. Depending on where we lived, I have different memories; her influences were part circumstance through what was available and part her own evolution in the kitchen. In Holland, she started making these mushroom vol-au-vents with a cream sauce that were insane. Back in Buenos Aires, she made the best *asados* on weekends; a rarity to see a woman at the helm of a *parrilla*, or grill, in Argentina. *Asados* are Argentina's most iconic culinary manifestation. It is our version of barbecue that takes place over many hours, usually on weekends, and is centred around the *parrilla*, which is unique to the Argentinian style of grilling meats over low heat for long periods of time, with no direct contact with the flames. There was also a squash gratin with cheese and a hint of added sweetness in the winter.

My father was and still is a great cook. While we did not live with him for most of our lives, the memories and influences are still there. The lightest potato gnocchi cooked in loads of burnt butter; once the gnocchi were gone, my brother and I would add more grated parmesan to the dish just to mop up the leftover butter. I learned to make mayonnaise with him, me standing on a stool drizzling the thinnest stream of oil until the sauce could take on more. My grandfather made even better *asados* than my mother (no wonder how she became so good) and I remember him starting the ceremony first thing in the morning with a kitchen towel thrown over his left shoulder as he salted all the meats. My grandmother Oma made the best apple cakes. Her German ancestry was all over her food and those apple cake recipes have been passed on from my mother to my daughter, while my son makes mayonnaise with me. It went beyond the food with Oma. Even for afternoon tea, she would set the table with an elegance and style that was pure celebration, but for her it was every day. I learned that the table setting is as important as the food.

Food is a deeply rich ingredient of our collective culture. As is conversation and daily family ritual; how and with whom we eat is just as important as what we eat. Immigrants around the world retain their cultural identity often not through language but through the food and ceremony around it. Many children of immigrant parents don't speak their parents' native language fluently, yet they grow up eating their native food. When we think about who we are, where we come from and what makes individuals from one village different to those of the neighbouring one, food is as fundamental as language. Sadly, in a similar fashion to our disconnect with nature, many of us have disconnected our identity from food.

As I embarked on my own cooking voyage, all these influences were a valuable foundation to build on and those vivid flavour memories were a guiding light. Alberto passed away in the Netherlands, and we quickly moved back to Milan where we lived for a couple of years before my mother decided we should move back to Buenos Aires. These years saw the beginning of my own cooking, where I realised the enjoyment and interest I got from it. When I graduated from high school, I went to university in Philadelphia and, as my first experience of living alone, I began to explore the food markets and started cooking for my friends. I remember going with Jeronimo, one of my closest friends from high school, to South Philly so we could buy tons of cured meats and cheese. The owner of the deli knew me well and always indulged me with copious amounts of charcuterie while I shopped. After graduating from university in 1996 and moving to New York City, I began to cook even more. I wanted to learn more techniques and become a better cook. I told myself I needed to master the basics: the best spaghetti pomodoro, the best roast chicken, a perfect *milanesa* where the breadcrumb coating on the veal does not bubble up when cooked so you get a beautifully crisp result (here, again, personal memories define perfection). The food I grew up with was in me and I wanted to perfect it, all of it, as a means of constructing a culinary framework that would enable my creativity to flourish. So that's what I did. I discovered Jacques Pépin's masterful book, *La Technique*, and worked cover to cover, learning all the classic techniques. Many I practised over and over. Interestingly, I realised I had absorbed many techniques through osmosis by watching my mother fold egg whites for mousse, or my dad's hand movements when making mayonnaise, and more. With technique came the search for ingredients.

The deeper I delved into cooking, the more the pleasure of eating, of flavour, that joy and happiness I had experienced as a child, led me on my own search. I could not settle for mediocre produce. I wanted to find those life-changing experiences, those magical ingredients that allow you to perfect a dish of such complex simplicity as spaghetti with tomato sauce. My enjoyment of food is so connected to my enjoyment of the time I have on this planet. I was fortunate my bar was set high early on in life; however, I was unaware that I would not stop until I could reach it and raise it further. I would turn my obsession with flavour into my life's work.

Flavour Collapse

Over time, I discovered that flavour is about more than joy and connection. What I stumbled on was an incredible validation of my obsession with flavour. So much of what is wrong with our food system can be explained through the lens of flavour, or rather the loss of it.

To fully appreciate the central role flavour plays in shaping our food system, it is essential to have a solid understanding of what exactly we eat today. How is most of the food we consume actually produced – both farmed and processed? What has happened to flavour and nutrition as a result of these changes? And how have we compensated for the loss in taste? How is flavour achieved now? It is certainly not coming straight from the field or paddock. I say achieved because since the 1950s there has been a steady decline in the flavour and nutritional density of fruits, vegetables, chicken, beef and almost all other primary ingredients. You'll notice that this loss of flavour coincides precisely with the rise of supermarkets, but it is not a coincidence – the two are intimately connected.

After the Second World War, our food system transformed dramatically with the rise of the supermarkets. A new dawn was upon us, with democracy rooted in the social contract of much of the Western world. As countries, societies and families were rebuilt, industries had to adapt to an economy now driven by consumer growth rather than war. How we grew and produced the food we ate, and where and how we consumed it underwent radical changes, bringing far-reaching consequences for the planet, our own health and the joy we derive from flavour.

Companies like Dow and DuPont had been producing vast amounts of chemicals to support the US government's war efforts. Nitrogen was used in the production of TNT and other explosives, which made their way into bombs. Many new plants were built for this purpose and, as the world moved towards a post-war economy, these companies had to find a way to retain the market share for their products. They found two core markets: agriculture and the consumer mass market.

At that time, farmers and scientists became aware of the importance of nutrients – primarily nitrogen, phosphorus and potassium – for crop vitality, which coincided with the need to sell huge quantities of these chemical elements. This was the beginning of the industrialisation of agriculture – specifically, high-input farming where we moved farmers onto a system of heavy tilling, chemical fertilisers and chemically derived pesticides.

During the war years women had to replace the workforce, and as the war ended, both men and women had to readjust to this new reality. Particularly in the US, this change was believed to have been significant enough to endure beyond the war. Families were sold the virtues of convenience, the idea that they were time strapped and needed industry to help them cope. From pre-portioned cake mixes to cling film, everything was geared towards making life easier. Consumers were offered a lifestyle with products that would make their home comfortable and desirable. With that came a disconnection from food and its preparation, and from the importance of seasonality, freshness, flavour and nutrition. It also, in the US at least, heralded the end of the family meal and with that the joy and cultural importance that comes with it.

Finally, the supermarket model, which traces its roots back to Piggly Wiggly, a self-service grocer that opened in 1916 in Memphis, Tennessee, began to really take hold. The 1950s was the decade when the supermarket model really solidified itself as the future of food shopping. Convenience became the priority and supermarkets provided that in abundance, offering a one-stop shop with huge choice and low prices. Through their offering and merchandising they leveraged the change in habits to drive the consumers towards more processed, more convenient food, taking the food system with it. It impacted the farming of raw ingredients as well as the manufacturing and heavy processing of those ingredients. Pricing, or 'value' as the supermarkets would have us believe, took on a central role in this new food distribution system, with two major consequences. The first acted as further fuel to the growth in processed and ready-made food, which was falsely perceived as being cheaper than making your own. The second, and perhaps more severe, consequence came from the downward price pressures that fed through the supply chain and ultimately onto farms, impacting the entire farming system: this was the birth of industrial farming.

This hijacking of our desire for convenience led to less cooking and increased consumption of processed food, which in turn drove a reduction in the range of fresh produce, coupled with year-round availability of what I classify as 'easy' produce. Sadly, most of the fresh produce on offer needs very little, if any, cooking. From my experience, the Western world consumes bananas, watermelons, apples, grapes, mangoes, avocados, oranges, grapefruits, lemons, limes, berries, tomatoes, cucumbers, peppers, courgettes, aubergines, carrots, potatoes and onions. They are all fruits except for carrots, onions and potatoes!

Why is that? Because all these ingredients are consumed raw or require very simple cooking. We think we cook but that's far from reality; we merely chop a very restricted selection of fruits and vegetables. Supermarkets have dumbed us down.

This loss of gastronomic knowledge – how to cook, seasonality, flavour expectations, eating together, even how to set a table – has left the consumer vulnerable because knowledge is power. As consumers moved their consumption from smaller specialised food stores to supermarkets, a consolidation of demand occurred that gave a very small number of buyers control over the food system and our diets. This was a grave mistake, as the food interests of humans around the world were handed over to a handful of institutions whose own interests were far removed from our own. It created a very powerful cycle, as most consumer demand is funnelled through the supermarkets; in the UK, supermarkets control 95 per cent of all food sold,[1] while in the US that figure stands at 85 per cent.[2]

This gave their buying teams the power to dictate what farmers should produce. For the last seventy years, a small group of individuals have decided what prices they're willing to pay, what characteristics they want their strawberries to have, what they prioritise (yield versus flavour) for pretty much everything we consume. I want to be very clear: collectively and unwittingly we relinquished the power of choice, the unparalleled power that consumer demand wields. The consequences have been tragic.

Farming had to accommodate the new demand, so, across the agricultural sector, things began to change. For simplicity, let's focus on scale as the driving force of this change. Scale was seen as the answer to the supermarkets' needs, coupled with the timely surge in an ever-growing mix of herbicides, pesticides and other tools, which the chemical companies were so eager to stimulate. The confluence of chemical companies, supermarkets and seed companies heralded a new age for the global food system.

As supermarkets grew, their logistics networks evolved with them. They also thrived on scale as a means of bringing more choice and lower prices. Over time, we moved from local farmers delivering into local stores to very large farms delivering into distribution centres that would feed thousands of stores. Now, along with price, shelf life became a priority. Demand in the farming system shifted to yield (lower prices) and shelf life (longer transportation) all at the expense of flavour and nutrition.

But it went much further. The adaptation of farming systems meant larger farms cultivating fewer varieties. Less biodiversity on land and in our diets. A farming system engineered to function in a chemical cycle of aggressive weed suppressant, followed by heavy fertiliser use and multiple pre-emptive and corrective pesticide applications, alongside intensive ploughing. All depleting soil fertility and resilience.

Then the seed companies got involved for the final insidious piece: the development of plant varieties that suited the supermarket buyers. A tomato plant that produces more tomatoes, does so faster, and whose fruit has thicker skins for long supply-chain journeys. Nothing comes for free in nature; when you select for yield or appearance, you almost always give up other valuable characteristics, most notably flavour.

Up until the dawn of supermarkets, we would select plant varieties primarily for flavour. Who the hell wouldn't? With thin skins that split because of the high sugar content – that's awesome. But not so for the supermarkets. We handed power to supermarket buyers and they, in turn, asked the farming community for produce that was cheap, long lasting and uniform.

They offered us strawberries for a few more months of the year – tasteless ones, but we bought them. So, they extended the season even further and we bought those as well. Now strawberries are available year-round, still tasteless, and we continue to buy them. They took one of the greatest pleasures on earth away from us: the joy of flavour.

Flavour's Wisdom

Flavour equals nutrition. The greatest validation in my career was discovering the indisputable connection between flavour and nutrition. When I realised that the main focus of our sourcing at Natoora, this obsession with seeking out amazing flavour, meant that we had been in effect sourcing for nutrition, it blew my mind. A new chapter for the business swiftly unfolded, while on a personal level it was a vindication of my single-minded focus on flavour.

The concept of 'nutritional wisdom' is fascinating and explains why we are designed to love food that tastes delicious. Flavour signals nutrition. Given that we need to eat to stay alive, flavour is the body's mechanism for seeking nutrition. Nutritional wisdom is the ability of the body and mind to use its senses of taste and smell, working together, to experience flavours and elicit emotional responses that make us eat what is good for us and stop eating once we've had enough. Fred Provenza, an emeritus professor at Utah State University, conducted studies on nutritional wisdom, demonstrating how animals can modify their diet based on deficiencies and needs. A goat facing an iron deficiency will feed on a plant that contains iron until that deficiency is normalised, at which point it will reduce or stop consuming that plant variety. The only reason why one peach tastes better to us than the other is down to its nutritional density.

There are numerous studies that have analysed nutritional decline. One of the most quoted is a study by the *British Food Journal* that compared fruits and vegetables grown in the 1930s with those grown in the 1980s. The results are alarming: calcium declined by 19 per cent, iron by 22 per cent and potassium by 14 per cent.[3] Further studies conducted in the late 1990s and early 2000s corroborated those findings.[4] In fact, these more recent studies uncovered that as far back as the 1940s, scientists had become aware of what they coined the 'dilution effect'. In essence, a strong correlation between the changes in farming practices and varieties grown since the Green Revolution at the turn of the twentieth century and the nutritional decline in our food. The dilution of nutrition.

Supermarkets have warped demand, seeking longer shelf life, year-round availability and cheaper food, which in turn has pushed breeders, seed companies and farmers to adapt to those demands. In *The Dorito Effect*, a book that specifically addresses the topic of nutrient loss, Mark Schatzker writes on crops: 'Just like chickens, they'd been selected to grow faster and bigger, and that was diluting nutrients. It was as though everything farmers had been doing for the last fifty years – breeding, fertilizing, spraying – was ganging up on nutrition.'[5] It is worth looking at what has happened with chickens as a parallel to the aforementioned fruit and vegetable studies, and because it is a shocking illustration of what is wrong with our food system. In the 1940s, a mature chicken ready for slaughter would be roughly twelve and up to seventeen weeks old, depending primarily on the breed and what the farmer was looking for. That's anywhere from over 80 to 120 days. Today's supermarket chickens reach maturity at a staggering 35 days.

But we managed to go further in our quest for yield: in just five weeks, chickens are 700 grams (50 per cent) heavier and need a third less food to reach that size. And price, you ask? The price of a chicken back in the 1950s was roughly five to ten times more than today. These 'advances' and 'cheap' food bring with them serious consequences because now that we have produced inert, tasteless protein, we must trick consumers into believing it is good for them. We do this by manipulating the way that tasteless chicken tastes. Most ready meals will be full of sugar, salt and flavourings, but even more sinister is what is known as 'enhanced' chicken, which has been injected with flavourings in liquid form to improve its flavour and moistness.

As we manipulate the genetics of a chicken in the quest for yield, allowing prices to come down, we lose nutrition. The nutritional density of a chicken directly correlates with what it ate, the environment it lived in and how much time it had to develop into an animal ready to be consumed. By treating food as the output of a production process like paperclips in a factory, rather than the living products they are, we created the industrial food system we live by. So, while chickens have been bred to be nutritionally deficient, our industrial food system is just as good at taking nutritious ingredients and, through excessive manipulation, processing nutrition out of them. A simple product like a hazelnut, if well farmed, will start life as a wholesome and highly complex ingredient rich in fats, amino acids, minerals, fibre and more. However, as the hazelnut undergoes processes like blitzing or blanching, nutrients are removed or destroyed. Vitamins, amino acids, minerals and fibre can all be altered or lost as we peel, chop, heat and blend ingredients, particularly when this is done in a factory environment. A McDonald's apple pie is an ultra-processed food, greatly lacking in nutrition and high in saturated fats and sugar, and bears no resemblance to a home-made apple pie. Home cooking is of course a form of processing, but we categorise it as 'minimally processed'.

While natural flavours are derived from plant or animal material and artificial flavours are synthesised, both are produced in a lab and in many cases will be chemically identical and just as unsafe. You expect a cherry-flavoured Gatorade to never have seen a real cherry, but you don't expect the drink marketed to improve your gut health to have flavour additives that are also not good for your gut.

The data is staggering, particularly in the most developed countries where the industrialisation of our food system is most advanced. Back in 2008, already 50.7 per cent of the food consumed in the UK was classed as ultra-processed, whereas in France and Italy it sat at 14.2 per cent and 13.4 per cent respectively.[6] A more recent study has seen a concerning increase in the UK to 63 per cent by 2018,[7] and the US data from that same year shows this figure to be 57 per cent for adults and a shocking 67 per cent for youths aged 2–19 years old.[8] All of this food has flavour added back into it through natural and artificial flavourings, sugars and salt, making it taste good to us yet lacking the nutrition our palates and minds perceive. It is the greatest scam in history: a fundamental tool we use for our survival and well-being that allows us to lead healthy lives has been sabotaged. And to top it all off, we do so by harming the only planet we have. In the words of Schatzker: 'The Dorito Effect, very simply, is what happens when food gets blander and flavor technology gets better.'[9]

I've always said that one of the beautiful things about nature is that we can't outsmart it. If a day comes when we are capable of farming salmon in half the time it takes nature, on nutrient-poor feed, pumping it with antibiotics and for it to taste better than wild salmon, we are all screwed. Until then, there is hope. The adage that 'if something is too good to be true, it probably is' is one of your best tools in making intelligent food choices.

My own experience, which I appreciate is not empirical or peer reviewed, confirms this. It is, without appreciating at the time the health and wider benefits, the reason I started Natoora – I was seeking lost flavour.

The Birth of an Idea

Natoora's roots go back to a moment imprinted in my head that took place in the winter of 1998 at the Citarella on 3rd Ave and 75th Street, New York. Citarella is a gourmet market, an NYC institution, and while it is known primarily for its seafood, you can find a great assortment of fresh food. It was here that my desire to make a change to the food system was ignited.

It was close to Christmas and snow was on the ground. I had just walked in, straight into the fresh produce aisles. There was a woman, in her late forties or early fifties, who asked one of the store staff where she could find the peaches. She was not asking if there were peaches; her expectation was that they were available, and she just needed help finding them. Instinctually, the striking thought I had was, why is her body craving peaches? To me, wanting peaches in the depths of winter was the equivalent of being out under the radiant sunshine on the deck of a sailing boat wearing nothing but my bathing suit and thinking I should cook a nice hearty stew for lunch.

The faults of the modern food distribution system were clear to me at the time, but that encounter at Citarella resonated with me because it was emotional. All at once, it encapsulated the de-education of consumers when it came to food. I kept asking myself: how have we reached a state where people are asking for peaches in December? It is as absurd as reaching for a goose-down jacket in the middle of August in New York, and yet with food we are made to feel it is normal. Food is one of the pillars of our human cultural heritage, the means and the knowledge to feed ourselves and stay alive, and here was proof that we were losing it.

When I moved to the US it was a radical adjustment from a food perspective. Even though I had attended American schools wherever we lived since the age of nine and had been exposed to the industrialised junk food that I so loved, thanks to my American schoolmates who had access to US military commissaries (essentially American supermarkets carrying everything from Tide laundry detergent to Hostess Snoballs, Gobstoppers and Kool-Aid), my move to the US amounted to a culture shock.

Nowadays, you can find Oreo cookies pretty much across the world. That wasn't the case when I was growing up; the only American brands you'd find outside the US were Coca-Cola, Pepsi and Marlboro. I was very well acquainted with American junk food, yet my move to the US was radical because of the difference in flavour and quality from what I was used to when it came to fresh produce. In the countries I grew up in, the standardisation of fruits and vegetables had not taken hold and the gastronomic culture was vibrant and localised, with the majority of people relying on daily cooking to feed their families.

What I didn't know then was that many years later, while living in another great city with similar issues, I would finally find the courage to devote my life to fixing this problem. I would figure out a way to give consumers living in these large cities access to great food. Having left a career in banking, and after much soul searching, I made the decision to turn my love of food into my work. In what at the time was a drastic change of career in all manner of ways, I ended up working for Solstice, a produce wholesaler that was supplying some of the best restaurants in London. I pitched the owner my idea – to set up a home-delivery service for consumers – that would solve the frustrations I first had back in New York. We would build an e-commerce platform that allowed consumers to order the same quality of food as the best chefs. Within a year, after building and launching the service, I took the decision to leave. And this is where the Natoora story begins.

Originally started in Paris by two brothers and a friend, Natoora had been brought over to London in 2004 by one of the original investors. The concept was exactly what I had built at Solstice, and I was asked to take over the nascent business in London. I took an existing structure and developed the Natoora we know today over the course of the months and years that followed.

Natoora, like Solstice before it, was my answer to a misguided need for peaches in December. Born as a solution for consumers to the flavour quagmire, Natoora evolved into a movement specialising in fruits and vegetables and encompassing the incredible restaurant ecosystem. This in turn led to a profound connection with the farming community around the world, nature and the crises facing our planet.

In the early years, the business was focused on serving the home cook, but all that changed when in 2006 we acquired another London-based home-delivery service called Portobello Food Company, which specialised in Italian food. The catch was that they also supplied a handful of top Italian restaurants in the city like Locanda Locatelli, Harry's Bar and, of course, The River Cafe. What I could never have imagined at the time is how important these restaurants would be to the future of Natoora.

A few years passed, and in April 2009 I made the single most important decision that defined and shaped the organisation's future. I decided that we would dedicate our energies to the sourcing and distribution of fresh fruits and vegetables. I cut the number of products we offered, focusing heavily on seasonal fruits and vegetables, and made the restaurant industry a priority. No longer would we think of home cooks as the core of our business; it was the chefs we wanted and needed to win over. These decisions set the foundations of our success.

Professionally, I knew flavour was the key to working with some of the world's most talented chefs – they, too, seek to create wonder through flavour. It was by working closely with Rose Gray and Ruth Rogers at The River Cafe that I realised I could increase the accessibility of great-tasting fruits and vegetables to both chefs and consumers, and bring seasonality back onto our plates.

We also began working with some of the most forward-thinking chefs in London, like Brett Graham from The Ledbury and James Lowe from Lyle's, who pushed us to go even further in our pursuit of flavour. Over time, we saw the need to foster the same community of farmers, chefs and consumers beyond London, and increase our impact from Los Angeles to Melbourne. We have built unique networks of over five hundred famers and producers (supply chains) bound by a common thread – that of bringing back amazing flavour while ensuring a sustainable future for the planet.

Federico Cervellin, who features heavily in the stories in this book and is both a friend and an ally in my search for flavour, joined Natoora in March 2009 at this critical juncture in our company's history. Barely into his twenties, Fede demonstrated a vast appetite for hard work combined with a sharp palate and a real eye for quality. From the start, it was evident that Fede had a connection with fresh produce that was unique: a gut feeling, a sixth sense; it was just a part of his soul. The language of food has no vocabulary; communication occurs with the sense of sight and that of touch. A love of food is one thing, but instinctive knowledge of fruits and vegetables is another – it is a rare, innate talent. In those early days, I brought him with me to New Covent Garden Market, London's main produce terminal, to do the night buying with me. This would be the beginning of a career that has seen him lead our product team for the last fifteen years. He has been as instrumental in building Natoora's incredible supply chain as I have, and many of our most innovative ideas have come from him. The story of those early years is for another book – a time of unconditional commitment and immense belief – but Fede's story and my own quest for flavour are inextricably linked. I am proud to be able to share a glimpse of his talent and commitment in the pages of this book.

I created Natoora out of a selfish need: a need to recapture magical moments from my childhood. Walking down the cobbled road in Positano on the Amalfi Coast to buy Sorrentine tomatoes, these gnarly fruits that tasted not just of summer but of that small stretch of coastline. One summer in Nerano, a fishing village on the Sorrentine Peninsula, our neighbour had us pick plum tomatoes that he grew opposite his house; the smell of tomato vine was so intense and even better was the taste of the *spaghetti al pomodoro* his wife made. These moments of pure simplicity, of impeccable culinary judgement, they can mark you forever. Natoora's mission to revolutionise the food system is an expression of that desire, of rekindling the joy of eating, the wonder of flavour.

Flavour as a Compass

I've been heavily influenced by Yvon Chouinard and the outdoor clothing company Patagonia that he founded back in 1973 – a company we deeply admire and respect at Natoora. Its mission is not only remarkable in its ambition but also given its genuine intent: 'We're in business to save our home planet.'

Our own evolution towards conscious capitalism was inspired by Yvon's great vision and the team at Patagonia.

There are a lot of parallels in how we engage and invest in our farming network and many of the innovations they implemented in their own supply chain, both at the farming and manufacturing level. I see Yvon and Patagonia as pioneers of the responsible and active approach to supply chains. Being engaged, understanding their humanity and finding creative ways to improve them. Reading in Yvon's book *Let My People Go Surfing* about Patagonia's decision to buy lockstitch sewing machines for their contractors and to work directly with cotton farmers, both in reaction to Yvon's decision in 1994 to switch all their cotton to organic within eighteen months, was at once inspiring and affirmational. As I said in my handwritten letter to Yvon back in 2014: 'Your work has given new meaning to mine and that of everyone at Natoora.' Four years later in May, off the back of my letter, I had the great fortune to have lunch at Patagonia headquarters in Ventura, California, with Yvon. His generosity to a fellow entrepreneur of a small, unknown company from England shows the stature of the man.

As time progressed, as the business and I matured, I realised we could do more. Much more. Time and time again, I have seen flavour vindicate itself as an indicator of positive outcomes. The correlation between farming practices, soil health, nutrition and flavour became more than apparent.

Recognising the importance of consumer demand into the food system was a turning point for me. Inspired by my microeconomics lectures in university and given my perspective from right smack in the middle of the food system, it was clear a demand-side shock was the necessary remedy to fix the food system. I felt I had a responsibility to act and demand change from consumers, to figure out a way to stimulate consumer demand. From here the concept of revolution took hold, of harnessing collective power to overturn a system and instil positive change.

Over the years, we've built an organisation united by an incredible global community that spans many of the great food cities of the world. I saw the opportunity beyond London, recognising that the value we bring to Noma in Copenhagen by virtue of the link to beautiful produce through our ever-growing farming community is the same we bring to Septime in Paris or Eleven Madison Park in New York. It is no coincidence that we work closely with Ilis restaurant in Brooklyn, which was opened by Mads Refslund, one of the founding chefs at Noma. Our community also includes committed consumers who believe in the power of flavour and its ability to stimulate demand for produce grown with integrity and mindfulness towards its externalities.

Natoora is the first of its kind. We are not the first company to bring farm-direct produce to restaurants, but we certainly are the first to do it at a scale never attempted before. If you are a chef in Melbourne and move to London or New York, not one of your fresh-product suppliers will have a presence in those cities, apart from Natoora. Operating at the highest level of quality, with a delicate, highly perishable product, sourced through a vast network of regional farms, is one hell of a challenge in one city, let alone around the world. As we've expanded beyond London, we've had to rebuild supply chains from the ground up – it has taken courage. As we can't ship freshly harvested fruits and vegetables from Sicily to Miami or Melbourne, the core of our work demands that we build a community of best-in-class farmers in each region we operate in.

This experience has given Fede and me the opportunity to visit farms across the globe, gaining valuable insight and perspective into the industry and specifically 'flavour-first' farming.

I have learned as much about food systems as I have on what it takes to scale a business that is inherently local into a global organisation. In an otherwise brandless space that is owned by supermarket own-label fruits and vegetables, we have created a brand occupying an entire category – again, an industry first anywhere in the world. Chefs and consumers alike can put their trust in our sourcing and ethics to deploy their purchasing power for meaningful impact.

The relevance of Natoora's unique and innovative journey is that I have gained a distinct perspective that gives me agency to discuss the problems and challenges, as well as provide what I believe are tangible solutions, to tackling the problems in our food landscape. Our food is overly processed, it has lost vast amounts of nutrition, and we are damaging our soils and planet as we produce food that is often harmful to our health. The way we grow, harvest, process and ultimately feed ourselves – what we call our food system – is broken. I believe I bring a new perspective on systemic change that is tangible, rooted in a life's work enacting real change. You'll read about winter tomatoes and green citrus, two of the most radically seasonal groups of products, that now, thanks to my work and that of Natoora, appear on menus across the world. We have inspired chefs and home cooks to look deeper into seasonality, to push and stretch the boundaries of flavour.

I know first-hand the limitations that are hardwired into the way our food gets from farm to plate, and how difficult it is to work within those limitations. But I have also found solutions. In chapter 6, I discuss what I call 'accessible scale' farms and how they are critical, within the context of a global food system, to a better future. Similarly, my exposure to the vast cultural differences around the world informs the stories that I share in this book.

My vantage point is based on real experience, and the farmers' stories I share help bring humanity and realism to the current situation and possible solutions to it. These stories are the same ones that pushed me to continue fighting for a better future. You'll feel the love and passion that is present in a terrific plum, and how valuing that, tasting that love, underpins our food choices. I trust these stories will not just serve to inform the arguments set forth but importantly inspire you too. As humans, we have grown through stories: they are as central and innate as fire is to our evolution as a species. Sharing these wonderful experiences of the people behind some of the best food being produced in the world today is more important than any single solution I can provide. My aim in writing this book is to provide hope and guidance: how our food choices, when driven by flavour, can leverage our collective power to rapidly change the food system.

The world has moved on from my childhood. We now hear about conscious consumption, about transparent supply chains, and how consumers want to know where and how products they buy have been produced. Consumers are demanding more – more information, more transparency, more accountability and a higher degree of ethical standards. The younger the consumer, the more ardent their belief in a system that is more just for all parties involved, including the planet.

Imagine if we could arm all our children with flavour wisdom, with the discerning palate to make intelligent choices for themselves, which in turn will deliver positive outcomes, stimulate the right demand and upend the system. Like a compass, flavour is invaluable in the absence of other data, it is the most valuable tool we have when deciding what we eat. Flavour holds the key to the future we need, to the food system of tomorrow.

A Hopeful Future

As I look back on the past two decades, I feel a great sense of hope. Like a climber standing at the foot of a great mountain they're about to ascend, I too feel the enormity of the challenge. Yet the mountains I've climbed so far, those I share in the pages that follow, provide me with tremendous hope. The human stories that are so central to everything I and everyone else at Natoora have done to connect beautifully farmed, flavour-first produce with chefs and consumers around the world, are perhaps the greatest source of hope.

What I know is that land is available. The varieties are certainly available. I can tell you that the farmers are far happier growing fruits and vegetables that taste amazing rather than what we've been asking of them. And then there's the planet. She will also be happier and will allow us and future generations to continue living in her wild natural paradise. The mountain can be summited.

I hope that by conveying a renewed sense of reverence for our food that more individuals like you will recapture that power to reshape our food system. I want us to reconnect with the joy of flavour and also with nature so that we collectively place the right value on healthy, nutritious food. By taking you to these magical places where we source our produce, you will understand how flavour is at the centre of revolutionising our food system. I want us to revalue beautiful food, to see the artistry and magic of it all. A green Tardivo di Ciaculli mandarin can alter your life forever, and it does so through flavour.

The beauty of mind-blowing flavour is that it creates a visceral reaction, an emotional connection that is terribly powerful. Joy cuts straight to the point. It enables the consumer to demand better. I fundamentally believe that the answer to fixing our food system is consumer demand. If we all stopped buying strawberries in the middle of winter, you can be certain that the supermarkets would stop selling them and the farmers wouldn't grow them. The more exposure we can create to quality food that tastes amazing, the greater the chances we have of shifting demand.

By raising the bar on what is possible when it comes to the taste and enjoyment of a peach, or even a single cherry, we arm the consumer with the desire to replicate those moments of joy. To elevate what is expected of our food system. In economic terms, to stimulate demand. The right kind of demand. Joy is more powerful, more enduring, than any sustainable metric.

Power needs to return to the consumer. A flavour-driven revolution.

CHAPTER ONE

Seasonality

Descartes' dictum 'I think, therefore I am' is humanity's defining statement of being. 'I eat, therefore I'm seasonal' should be humanity's defining statement of eating.

Our earth's seasons are a gift from the Big Bang, our solar system or some kind soul out in space. The angle of the earth's tilt on its central axis is responsible for the changing weather over the course of the year, all 365 days, and whatever caused it we should be grateful; for without this tilt, we would have no seasons, only day and night. We inhabit this earth, a magical place with an environment devoted to the flourishing of life. We are a product of it: we do not stand outside of nature; we're a mere creation of hers. Understanding our planet and remembering our connection to nature is essential if we are to see the seasons as a precursor to us.

That was the norm until only a handful of decades ago when modernisation, improved transport, technology, consumer habits and the mass distribution of food eliminated natural seasons to the point that today eating produce that is in its natural season is incredibly difficult. Consumers have lost all knowledge of seasonality, as shown by a BBC report from 2014 that found less than 10 per cent of people in the UK knew when some of the most popular fruits were in season.[1] Yet there is great joy in letting nature dictate what we eat. Produce that is in season tastes better and is less expensive, and there is something deeper to eating seasonally.

The Philosophy of Eating

Allowing the seasons to dictate what we cook opens a deeper connection to nature and its rhythms. This is a healthy relationship to foster. One of the reasons why we have ended up where we are is the disconnect with nature, and by removing ourselves from its rhythm we risk permanently damaging the ecosystems that sustain us on planet Earth.

True seasonal eating is filled with joy. Unavailability and scarcity breed resourcefulness in the kitchen, and you begin to discover a wider range of produce and interesting combinations. Your cooking evolves, becomes more natural and less prescriptive – it simply gets better. Simpler. You begin looking at produce more closely and gain a deeper understanding of how a sprouting broccoli evolves through its season, so you adapt your cooking to it. It is how instinct in the kitchen is developed. These are the real reasons we should all be religiously seasonal in our food choices, if only it were easier to achieve.

Seasonality has been co-opted and distorted, used as a buzzword to entice us. You walk into a healthy salad-bar chain with 'seasonal healthy food' displayed on the window, yet you find cherry tomatoes on offer all year. This has only made it harder to eat within the boundaries of the natural seasons. The infrastructure – the way we design menus, our seasonal education, even our own eating habits – needs to change and evolve.

Menus, particularly those of larger restaurant groups or chains, need to evolve as they are stuck in the past. The menu-design process, along with customer expectations – wanting the same dishes on the menu every time we visit – need wholesale change. There is a beautiful parallel with the redesign concept and how it is used within the context of the circular economy. For a product to be circular, it needs to be recycled and to not end up in the junk pile at the end of its life. The way to achieve this is to build circularity (recyclability) into the product, and therefore we need to redesign many of our everyday products with this in mind. The menu of that healthy salad chain is conditioned to only change twice a year, perhaps it even changes monthly, but it must cater to customers who demand tomatoes year round. The same is true of the products we find on supermarket shelves. I believe the way to overcome these constraints is through process redesign, just like redesigning the motor of a car so reusability or recyclability is inbuilt.

We developed a range of dips, sauces and soups at Natoora, initially for sale in our London stores. The idea was simple, we thought. Use exclusively ingredients in season, of the highest quality, and be fully transparent in their sourcing. From this exercise, the concept of 'radical seasonality' came about – pushing the envelope and the mindset of what is a truly seasonal product. Initially, we swapped and changed recipes as the seasons progressed but, as we put product into larger distribution networks, we came face to face with the constraints. Supermarkets give you a listing, a place on the shelf, and losing that place is not commercially smart. They also don't want to remove a product that sells well, and that is why the fresh dips on supermarket shelves are there year round, irrespective of the season. And why someone, somewhere, is growing those red peppers and making them available to the dip producer every day of the year. We faced this precise challenge, but the difference is that we tackled these challenges by reducing the range and creating winter and summer products. Conscious that our prepared foods could positively educate consumers, we were adamant we would not use peppers, a summer ingredient, in the winter. The solution? Our muhammara, a Middle Eastern pepper and walnut dip, is made with fresh peppers in the summer, then we switch the recipe to use dried choricero peppers in the winter. Continuity of line, which the supermarkets want, with radical seasonality so we stay true to our mission. It is about finding a level of compromise that works so the system can shift towards more seasonal eating.

Plants are living beings and, just like us, thrive when life is good. It is simple: a thriving plant becomes a source of great food. By consuming produce in season, we are eating plants that were living their ideal life, which is both sensible and, importantly, maximises flavour. We can go one level deeper into the meaning of eating seasonally. Submit to nature and entrust our fresh food choices to it. This is her culinary education. To do so, we need to run a finer comb through seasonality.

Seasonality 365

As a consumer, if you're not close to seasonality every day it is easy to lose track of it very quickly, and many of the organisations that profess to be seasonal are not. This is especially true of the chain restaurants providing an alternative to traditional fast-food outlets, and I believe this perpetuates the confusion.

Winter, spring, summer and autumn have provided us with logical divisions of the calendar year; however, they are a blunt, outdated tool. I want to apply the concept of microseasons and take it further – to seasonality as a 365-day continuum. The four seasons give us a framework based on scientific and astrological realities, yet seasonality is a day-by-day process that repeats itself every calendar year. I want to engage people in understanding this because it helps us to further develop our relationship with food; something that has been lost in more than half a century of industrialisation and the dumbing down of food.

For when you understand why a Tarocco Gallo orange grown at the foot of Mount Etna on mineral-rich clay soil is at its peak from late January or early February to mid-to-late March, you begin to understand how best to eat it, how best to cook with it and what best to marry it with.

You understand why your body, a perennial of the animal world, should also be craving peaches in late summer or a chicory plant in the middle of winter, alongside other nutrient-rich foods like fish and dairy. This is precisely why one should not be craving peaches in the middle of winter; our body should not be clamouring for them.

Our bodies are rooted in the natural world and we are just as biological as the fish, plants and trees that cohabit our earth. Our bodies' lightning-fast adaptability to high-speed travel is responsible for re-engaging with summer at the first breath of densely warm air as you disembark from a long-haul flight down the clattering aluminium stairs. Yet one should not confuse adaptability with dependency. Every living creature on this earth, from amoebas to humans, is deeply connected to nature's cycles and with the 365-day cycle we call a year, when our planet completes its tour around the sun. (Strictly speaking, it takes Earth about 365.25 days to orbit the sun, which is why we have leap years, but our seasons aren't concerned with such technicalities!)

By thinking within this theoretical 365-day seasonal process, it forces us to engage with varietals. Plant varieties can be thought of as subcategories of plant species. There are thousands of varieties of orange trees, so calling an orange-coloured round piece of fruit an 'orange' is correct but as specific as calling a four-wheeled motorised vehicle a car. It is more specific to identify the car as a 'roadster', a two-seater convertible, and even more precise to identify it as a Mazda MX-5. The same is true of the orange. A Tarocco Gallo, one variety of the cultivar Tarocco orange, from the foothills of Mount Etna, is more exact, and getting to know the specific varieties of fruits and vegetables we consume is fundamental if we are to extract maximum enjoyment.

Similarly, not all apples are the same. Some are very tart while others are overbearingly sweet. When fully ripe, a certain variety's flesh will hold up well to cooking while others collapse when cooked. Without this knowledge it is very hard to make the most of the food we eat. There are roughly 7,500 eating apple varieties that we know of around the world, yet only about 10 are readily available in supermarkets.

Along any timeline, say a month, distinct varietals are coming in and out of season. Apples are traditionally an autumn crop, but, like most tree fruit, a tree of a singular variety will be fully ripe in the space of one or two weeks. If you have an apple orchard with a consistent microclimate all planted with Granny Smith apples, you will only have freshly harvested apples for a couple of weeks. For this reason, most fruit orchards, from peaches to oranges, are planted with several different varieties, all of which have different ripening times, thus allowing a farmer to have peaches available for three months rather than two weeks.

Meander through Carmelo's citrus orchards at the foot of Mount Etna in Scordia, arguably one of the finest citrus-growing regions in the world, and you'll see different subspecies of each core varietal. We met Carmelo many years ago as our sourcing took us into the famous Sicilian citrus groves, and over time and a shared love of great food, including sea urchins and red prawns, we have built a wonderful friendship. The Tarocco orange is one species of orange defined by its red-tinted flesh and skin. These oranges are more commonly bracketed under the blood orange name, which derives from a different species of orange, also grown in Sicily, known as the *Sanguinello*; *sangue* in Italian is the word for blood. Carmelo grows four different varieties of Taroccos, each a slightly different varietal within the Tarocco family, and each with a different ripening timeframe. So, even though oranges are traditionally a winter fruit, if we refine our view of seasonality, we will see within his orchard an orange season that begins, depending on the weather that year, in late autumn or early winter and ends in the middle of spring.

If you take a two- or four-week snapshot in that instant, you have pure seasonality, no longer dictated by winter or spring but rather by the products that cohabit their peak around the same time. Being more precise is educational, knowledge is power, and when you combine varietal knowledge with plant life cycles you see why a 'winter' fruit like oranges goes so well with the young tops of the *sanapone* plant, a 'spring' mustard green.

Early, Peak, Late

I've always loved the concept of 'early, peak and late' as an improved way of communicating varietal-specific information to our customers. I was inspired by the Japanese, who apply a great level of detail to seasonality. In Japanese culture, a turnip will have a different name depending on whether the turnip is in the early part of its natural season or late in its cycle. Each plant, and within that each variety, has its own seasonality and life cycle and moves through a period of early, peak and late that is its own.

To illustrate, I will generalise. Fruit tends to have a very defined early period where sugars are lower and acidity is higher, with generally a sharp fall-off during its late phase where quality degrades substantially in both flavour and texture. On the other hand, leafy greens tend to have very short and less defined early periods, moving very quickly into a peak state before they gradually fall off into late with a less pronounced degradation than fruit.

For many years at Natoora, we have been using early, peak and late (EPL) as the key system to categorise all our produce, fitting in nicely with the concept of 'seasonality 365'. It is a fundamental tool that helps chefs build more seasonal menus by staying on top of changes at a product-by-product level. I have heard first-hand how this new language has shifted the way kitchens approach their menu writing, particularly those restaurants who have greater flexibility in changing them. Seasonality is very complex nowadays, even for professionals. If you're not in the thick of it, remembering when each variety is at its best is a tough ask.

EPL does a lot of the heavy lifting for chefs and consumers. It is the driving force to integrate seasonality into our lives and onto restaurant menus, giving us a greater appreciation for the nuances of a parsnip across a season. When are sugars more pronounced and why, how does its flavour and texture change as it moves from early in its season to peak and therefore how does my cooking evolve with it? Language is a valuable tool in changing the way we look at seasonality in a very fundamental way. No longer is a turnip in season or out of season, but rather how does that turnip flow through a defined timeframe alongside other plants? Language, and particularly that of EPL, serves as an educational instrument.

The beauty derived by this appreciation of a plant's life cycle is enormous, as it moves us closer to nature and allows us a deeper sense of feeling and understanding. It creates a different pace, slowing time down, and in doing so elevates the appreciation we have for the food we consume. Knowledge, once again, is power, but more so than power, it is a tool to extract more joy from life.

The Impact of Seasonal Cooking

It's not down to chance that peas and broad beans go well together, or that radicchios and oranges complement each other fascinatingly; what is even more telling is that late-winter produce, like oranges, go beautifully with the first broad beans. The overlap of winter to spring demonstrates that at those seasonal extremities there is another sub-season, where those products are actually of the same time and therefore can sit in harmony on our plates. This is how we derive real pleasure from food, by understanding these relationships so we can create magical flavours. We need to re-educate ourselves and return to our bodies the natural, instinctive flow with which we crave our foods.

Eating seasonally is not a trend, a catchphrase or the latest superfood gimmick; it is the most natural and sensible way for humans to eat. It is imbued with logic and common sense and should dictate what we decide to eat. Rediscovering the seasons is enlightening and revealing. It brings pure joy and awakens an understanding of our place on this earth and the prevailing strength and power of nature; a nature that created us as well.

By combining a deeper sense of the seasons with the specific varieties of produce we consume, we will demand more from our food. We will ask our food system for more biodiversity, for produce in its natural season and, as we do, the system will restore itself, putting flavour, nutrition, and the health of our soils and our planet front and centre. We as consumers hold the key to revolutionising the food system.

I eat, therefore I'm seasonal.

CHAPTER TWO

It Starts with a Seed

Our connection to the land dates back over hundreds of thousands of years, to a time when we roamed the planet as hunter-gatherers and allowed nature to feed us with little to no intervention. Then, around twelve thousand years ago, the single most important development in human history occurred: we invented agriculture, allowing humanity to evolve from roaming communities to modern, complex societies. As we built a self-sustaining food production system, it allowed us to settle, feed larger communities and specialise. The specialisation of labour gave the human race a great leap – without agriculture we would not fly, there would be no computers or artificial intelligence and no nuclear weapons – and it all started with a seed.

If you've ever held a seed in the palm of your hand, perhaps you will have wondered how life can be stored in such a small package. Everything a plant needs is stored inside its seed coat, awaiting the right conditions to give birth to a new plant. All the traits and peculiarities are nestled into its genetic code, predetermined, predefined. Only mother nature could create a reproductive system of such bewildering ingenuity. The combination of human intervention, the act of intentionally planting a seed, with nature and its laws of reproduction, is what gave birth to the development, or better said the furthering, of flavour. There is no question that flavour's journey begins with a seed.

Selecting for Flavour

Without getting into an overly complex analysis of plant physiology and reproduction, there are two ways in which humans have developed the fruits and vegetables we consume today. The first is by selection and the second is through breeding. Initially, to increase the quantity and predictability of available food, our ancestors collected seeds from wild plants. As those seeds produced offspring that were similar or identical to the wild parent, the first farmers selectively saved seed from the best-performing specimens and plant selection was born. Plant species evolve, as all living organisms on the planet do, by mating and through natural selection: roots get longer, leaves go a darker shade of green, some are stronger and resist disease, fruits sweeten.

Plant breeding takes selection one step further by actively crossing or mating two plants to create a seed that will produce a plant with specific characteristics. Breeding is what has given us citrus fruits like the Tacle, which is a cross between an orange and a clementine. One of the great examples of modern plant breeding is the sugar snap pea. The late Calvin Lamborn was tasked in 1969 with breeding a straighter, smoother snow pea, what is also known as the mangetout. During his breeding programme, he noticed a pea with a thicker than normal pod and wondered whether this peculiar characteristic could yield a straighter pod. So, he selected this plant and used it to cross-breed with more normal-looking snow peas. Roughly a decade of progressive breeding led to a completely new product that is now consumed around the world: the 'sugar snap'. A pea that looks like a garden pea, also known as an English pea, yet with an entirely edible pod. I've had the fortune of working with his son Rod and exploring many incredible pea varieties that the family has developed over the years. (Quite auspiciously, Gregor Mendel, the Austrian monk who is regarded as the father of genetics, did most of his experimentation using the pea plant.)

Pea plants are self-pollinating and, more specifically, the flowers have both male and female parts. By controlling pollination through manual intervention, it is possible to cross-breed – to take pollen from one plant with specific characteristics like a thicker pod and pollinate a flower with different yet desirable characteristics. That flower will now produce a 'new' pea variety carrying genetic material from both parents. Over ten thousand years of human intervention, the careful selection or intentional cross-breeding of plants has created an unquantifiable richness in our food system and is an immensely valuable piece of our human heritage, one which needs preserving.

What we choose to select or breed defines what we eat and, importantly, how food tastes and the nutrition within it. We can divide the history of plant breeding into two, or perhaps that is how I choose to divide it. A flavour era of agriculture, which dates back from twelve thousand years ago to the 1950s, and a yield era of the past eighty years. In human terms, only six months of an entire life would account for the years where flavour was not at the forefront of agriculture, yet we have undone so much of the past 11,420 years of human innovation. Prior to the 1950s, there was a beautiful feedback loop at play. In our quest to satisfy pleasure through flavour, we enhanced nutrition.

The Importance of Varieties

The seed or, to be more specific, the variety is the single most important contributor to flavour. I know this is controversial because most literature out there credits the soil, but I do not agree. Just like a chef is only as good as the produce she starts with, a farmer is only as good as the variety she plants. And like a chef, a farmer can destroy a great variety or enhance it further through good farming practices and, importantly, experience and expertise. The more lemon tarts a chef makes, the more she perfects all aspects of it and in the process maximises flavour. The same is true of farming, and it explains why one farmer's chicories are superior to another's. Soil is, of course, extremely important, but it cannot compensate for poor genetics. Take two identical seeds, plant them in soils with different levels of organic matter and microbiology, and, yes, the healthier soil will deliver more flavour and nutrition. If you were to plant your typical supermarket green courgette variety developed for shelf life and yield in great soil, you will not magically produce a good-tasting courgette; whereas a courgette variety bred for flavour but that is cultivated in poor soil will undoubtedly still have the superior taste.

As a consumer, you have nature on your side. Getting to know specific varieties is really valuable for a number of reasons. Beyond knowing that a melon variety is super tasty, there are, you could say, some inbuilt protections in your choice – a sort of natural selection at play without needing to know anything else about how or who farmed that melon. If we go back to preagricultural times, those wild plants that we domesticated were naturally adapted to the region they were growing in. Take a coastal plant from Scotland and it is unlikely to survive in the French Alps – it would certainly be terribly confused if the seed managed to germinate and set roots. This locality, which is built into the genetics of the seed, gives you a level of protection because it mandates farming within a specific geographic area. This is the cornerstone of the PDO (protected designation of origin) certification system that guarantees unique origins and product authenticity. Additionally, in the same way a race car has not been designed to be driven offroad, an old melon variety will not be suited to a highly intensive farming regime. Varieties hold huge benefits in protecting quality through their genetics, as well as how and where they are farmed. So, when we seek out an indigenous variety, we are 'buying into' not only an ecotype with distinct flavour and textural characteristics, but also a particular microclimate, soil type and culture that exalt it.

Olive oil immediately comes to mind as one of the products that is most iconic in this respect. Monocultivar oils are produced from one single variety of olive, as opposed to the majority of olive oils out there, which are blends. Like wine, when you bottle a single grape variety like a Pinot Noir, you get the full expression of the varietal as opposed to mixing it with other varieties. As far as fruits go, the olive is something of a prodigy of the plant world, whose ability to release one of the most awesome ingredients takes it straight to the top of the edible plant kingdom. Much of the world's cooking, of our human heritage, would not be the same without the olive. Like salt, it is one of the great pillars of our civilisation, a dignified conqueror of vast swathes of Mediterranean land.

There is an olive oil at Natoora that goes by the name Senia, which I believe is one of the best in the world. And its secret lies in the single variety of olive used, the mighty Tonda Iblea. It has become an institution at Natoora, from Melbourne to Miami.

I discovered it thanks to the sublime respect, the reverence for ingredients, at what is hands down my favourite restaurant in the world. Sitting quietly over the water's edge with a view of the great Etna volcano, it is fitting that Carmelo, when I first visited his orange orchards in July 2012, brought me here for dinner. In the little village of Brucoli, south of Catania, Emanuele and Viviana opened I Rizzari restaurant in what feels like a slice of paradise. A rare gem in today's fast-paced restaurant world, it pairs impeccable sourcing with an absolute simplicity that can only be the product of a deep connection to specific varieties of fish, vegetables and even oils. What Emanuele does with fish and a couple of ingredients is poetry. A kind and generous heart shared what others would have guarded. Plate after plate, this alluring oil kept appearing and disappearing. The cooking allowed it to shine and led me to ask what oil it was. Before the night was over, I was presented with a little unopened tin of Senia. On this tin was a phone number, no name, nothing more.

Days later, I was in the olive groves, surrounded by Tonda Iblea trees, a varietal that is unassailably considered one of the great olive varieties of the world. Its perfume is sensational, with green tomato and freshly cut spiky artichoke aromas flying out of its surface. The growing epicentre of this unique varietal is Chiaramonte Gulfi, whose soil and microclimate are home to the finest Tonda Iblea olives; that little nook of Sicily is truly special. The Tonda Iblea can trace its history back to the eighth or ninth century BCE, having been brought to the island by the Phoenicians and Greeks. But it was not until the end of the eighteenth century that one can see evidence of commercial cultivation of the Tonda Iblea in Chiaramonte tied to a decree that opened up land for agriculture. Its prominent place in the olive world was sealed when, in the 1970s, one of the most respected gastronomic voices in Italy, Luigi Veronelli, claimed that Tonda Iblea oil was the best in Sicily.

A serendipitous moment brought us here and we found an equally serendipitous and marvellously passionate man, Artilio, who loved showing us around, telling us about his trees, the soil, his father-in-law who owned the land, and his son Daniele and wife Claudia, who were moving from Catania to tend to the orchard full time. Again, like grapes where some varieties are ideal for making wine while others are best eaten off the vine, some olive varieties are great for oil while others are best preserved. Boy, do these olives want to be converted into oil.

Small, bright pale green, and ever so slightly round with a little pointed tip, it is another example of flavour over yield. Squish open a green olive, a rare sight after harvest, and the perfume of its flesh is almost identical to that of the oil we tasted an hour ago. There is an utterly pure correlation between the two, with no loss of flavour from fruit to oil. The ability of an olive tree to convert water into fruit into olives on its branches is extremely important in defining the final flavour of the oil – the better its ability, the more diluted the flavour. And this ability is defined by the variety – that is, by the seed. A Tonda Iblea is not as good at converting water into olives as the Picual variety, which produces an oil of far less flavour complexity and quality but yields a lot more oil per kilogram of olives. The seed, and therefore the varietal, holds all the starting blocks of flavour, or lack thereof.

The olives are harvested over the course of thirty days, non-stop, each day a process that takes close to fifteen hours from start to finish. Daniele starts with trees bearing one of the best olive varieties in the world, in a parcel of land perfectly in tune with that variety. Farming with care and flavour in mind, he harvests with the same ethos, very early at great cost to yield. When you harvest in late September, as opposed to November for example, not only is your fruit smaller but its yield is lower as the amount of oil present per olive is lower. You are compensated greatly, however, in flavour and complexity, and the overall quality of the oil skyrockets. Yields can vary from as low as 9 per cent to 12 per cent for the Tonda Iblea, which is far lower than most.

Daniele is a young man; his spirit is strong and as a very talented sailor his connection to the sea is fundamental to his work now as an olive famer. Sailors have a natural affinity with nature and understand more than most the respect it commands, for when you've been face to face with a terrifying storm, you learn to respect the power that nature wields. I am convinced that this connection to nature helps Daniele care for his trees, protect the land and soil they live on, and nurtures his humility to let nature be the guide that results in an incredible quality of olive oil.

There is peace and tranquillity here, and a gigantic carob tree with an enormous cavity formed by large arching branches, so big that you can enter it. The earth beneath is dry, covered as in the orange groves by a vast, infinite carpet of trefoil-leafed sorrel. Young trees give way to larger, older trees – grandfathers, some over one hundred years old. The gnarly Tonda Iblea trunks are life-like, full of charisma and scars, weathered. Even though Daniele does everything in his power to coax the best out of his olive trees, if his trees produced the Picual olive, rather than the Tonda Iblea, his oil would not be anywhere near as good.

Durability over Distance

Rather than developing produce so that it tastes better, so that it is more nutritious, we've treated seed development as if we were building inert products. As we consume more and more of our food from supermarkets, we have started favouring supply chains (distribution systems) that are longer and more centralised, where products have to travel further from farm to store, and therefore the ability of a lemon to travel longer distances is favoured over how great its juice tastes.

Think about this approach and how incongruous it is. We have to make food sturdier, cheaper, easier to harvest, better suited to sit on a boat for three weeks, less prone to disease, more uniform. This is OK for manufacturing screws – we make them tougher, more versatile, cheaper, lighter, identical – but this should never happen with our food. The technology that nature has created to allow plants to reproduce and evolve – the seed – has been sabotaged by an industry that today controls most of the food we consume. Simply put, the move to buying our food through the supermarkets gave their buyers the power to request non-flavour-driven attributes from varieties at flavour's expense. We traded thick skins for taste. We engineered the product to fit the distribution system, not the other way around.

Unfortunately, flavour-first plants are usually harder to grow, and they are also harder to transport. More effort is required to work with these fruits and vegetables, but it is possible to do so at scale; I know this first-hand. Until consumers demand better, though, supermarkets will not be incentivised to do the heavy lifting required to adapt their supply chains. And the necessary change needs to come in partnership with supermarkets as well as the large food buyers and food manufacturers.

The Cuore di Sorrento Tomato

Our breeding capabilities were not all put to bad use, and we should not view all modern varieties as inferior to what our grandparents farmed. No doubt a quick and easy way to revert to better-tasting food is to find those varieties that have only ever been primarily selected to taste amazing, but there is a lot of good innovation that has created great new varieties. Some of the best-tasting stone fruits, namely peaches and nectarines, are the product of varieties developed in this century. Some of the best-tasting tomatoes grown in winter are the result of innovation in farming techniques dating back to the 1970s; these are all very recent advances.

South of Naples, the Sorrentine Peninsula stretches west into the Mediterranean Sea and forms on its northern side part of the Gulf of Naples, while on its southern side lies the Amalfi Coast, an incredibly stunning, magical, romantic place. It is one of the most naturally beautiful pieces of coastline anywhere in the world. The crystalline multihued waters of the sea crash into steep, rocky cliffs covered in lush vegetation. Here, a mountainous green earth element comes into direct, almost mythical contact with a great expanse of water, with the towering sun as fire. Nowhere else on earth have I seen elemental beauty of this scale.

Volcanoes, sea and sun combine to provide an ideal growing environment for certain fruits, none more so than tomatoes. The mineral complexities of volcanic soil, the dry, rich sea air, the high-salinity water and abundant sunshine provide tomato plants with a perfect environment for them to thrive in.

I discovered the Cuore di Sorrento tomato as a child; my mother's love of the Amalfi Coast brought us to Positano during the summer months and, along with lemon granitas, tomatoes were an almost daily ritual. To this day, in the summer months locals eat almost exclusively the Cuore di Sorrento variety; it is autochthonous to its land, thrives here and is rarely known anywhere else. Back in 2011, we imported the first Cuore di Sorrento tomatoes into the UK, but it was not until 2014 that we found Raffaele, a young, eminently committed grower based in Torre del Greco at the foot of Mount Vesuvius. This rich piece of land looks out to the marvel that is the island of Capri.

To be truthful with our customers and distinguish a Cuore di Sorrento grown in its original birthplace from the area where Raffaele grows his, which is further north along the Gulf of Naples, Federico and I decided to rename the tomato Cuore del Vesuvio. There is a peculiar microclimate here, south of the town of Torre del Greco and roughly three kilometres north of

Torre Annunziata. Next to Sicily, it is one of the areas in Italy with the least amount of rainfall, and humidity here is 30 to 40 per cent less than outside this tiny parcel of land – this is how 'micro' the climate is. Winters are mild while summers are very hot with an abundance of light throughout the year, which is fundamental to growing good tomatoes.

No one knows precisely how old the Sorrento variety is, but what is known is that come summer in this part of the world, this is what families grew. They all kept their own seed, each family selecting and breeding their own specific ecotype, all cousins of the same original variety. You can see flavour-first selection at work, from the outer physical characteristics, the most notable of which is the extremely common scarring at the top of the fruit. Scarring is brought on by a concentration of sugars and the intensity of the sun's rays, which together reduce the elasticity of the tomato's skin. Importantly, the denser the tomato internally, the more prone it is to split. All of these characteristics have been bred out of modern tomatoes. Thin skins split easily, and where those scars form mould can develop. This is exactly what supermarkets don't want, but flavour is precisely what families were after and they couldn't care less about some scarring; in fact, they are a sign of beauty and quality, and characteristic of this variety.

Cuore di Sorrento's true season is late July through to the first weeks in October, yet Raffaele manages around eight months from the end of April all the way to December by utilising greenhouse cultivation and years of experience to straddle the traditional outdoor growing season. At Natoora, we source for just over four of those months, from mid to late May up until the beginning of October, as we like to stick as close to the old-school season as is sensible while retaining some flexibility to provide continuity for what is the absolute pinnacle of tomatoes, every bit as imposing, romantic, refined and utterly stunning as the Amalfi Coast.

In today's world of industrially manufactured tomatoes, those that are hydroponically grown and have never been in contact with soil, from modern seed bred for everything but flavour, are freaks of nature – more a testament to human intervention, manipulation and technological prowess. In contrast, the Sorrentos are a grass-roots revolution to the core; they are nature's ancient monuments left for future generations to bask in their magnificence. Quirky, lopsided, oblong and pear-shaped, the Sorrento has a pinkish skin and flesh, which ripens to a purplish complexion; its red is always tinged with hues of blue. Early in the season, it carries beautiful green striations emerging from a deep green top, which later it loses as it gains in flavour, oddly so. These characteristics are all present in the seed; the palette of colours Raffaele is handed at the start of each season. He can make a mess or paint a masterpiece. But it is that original choice of palette that is the biggest contributor to flavour.

Visually, a fruit or a vegetable can tell you so much about its quality – it is plain ignorance to argue otherwise – though the rest remains inside. There are two ways to investigate a tomato and both will yield valuable insight: vertically and horizontally. The vertical cut will reveal the seed clusters; in Sorrentos these are compact, displaying a dense, bumpy flesh with a shiny patina, with little gelatinous matter around the seeds. These clusters do not go through from top to bottom as you find in most standard round tomatoes but rather stop around two-thirds down, meaning the bottom third of the tomato is pure, unadulterated flesh. Due to this trait, Sorrento tomatoes tend to be heavier than similarly sized ones of a different variety.

A clean horizontal cut across a Cuore di Sorrento will reveal a percentage of flesh to seed that is the highest I've ever seen.

Meaty – a tomato steak – it is dense, rich, fleshy, ballsy; it is all about the mass, which, when blessed by fantastic olive oil, illustrates perfectly why mozzarella is the product of the same microculture. This fleshy characteristic is perhaps the most iconic of the Sorrento tomato and, in my opinion, the most important of various traits that make this tomato one of the best in the world. I am, of course, biased, by virtue of my childhood memories.

The story of this variety can help shed some light onto what is needed at the supply chain level to move the food system forward with flavour at its core. Sorrento tomatoes should be consumed fully ripe, properly red, intense and at their peak, but in line with the displacement of flavour as the main driver of quality, local markets want the Sorrentos to arrive on the green side with only a hint of pink coming through. They do this because consumers have changed appetite; they no longer want tomatoes that have been ripened on the vine, that bruise easily and have a very short shelf life once they bring them home. Back in the day, we were accustomed to dealing with ripe produce at home, cutting away mouldy edges, turning tomatoes into sauce, not afraid of working with the best that nature can offer. Even in Italy, consumers are seeking this incorrect vision of 'perfection' in our produce. I believe strongly in innovation, as strongly as I believe in history and tradition, and it is my view that in our industry you have to straddle both in order to succeed. Innovation was used to the detriment of heritage, to the detriment of nutrition, flavour and quality. It has fuelled a progressive deterioration of the joy within our food system and with it a deterioration of the planet as our aggressive farming practices leave no room for restoration.

We can turn the model on its head and use innovation to elevate and redirect growing traditions to produce outstanding flavour, not shelf life or unnatural appearance – perfection is not uniformity; perfection is found in imperfection. I know it can work, as it is precisely what we have been doing at Natoora for close to two decades.

It requires foresight and dedication on behalf of both parties and buy-in from the whole supply chain, which includes the end consumer. At Natoora, our sourcing innovation is profound and bred into our DNA, and I believe this is something that can and must be replicated by others in the industry. We've built technology to facilitate broadening the number of farms we buy one single product from, going completely against the industry norm of rationalising the number of farms to make it easier and more efficient. Green citrus, which I get into in Creating a Market for Green Citrus on page 121, is another great example of innovation: harvesting fruit when technically unripe and working with chefs to develop a new category of produce.

We spent many years fine-tuning ripeness with Raffaele, dedicating a lot of time on both sides to getting it right. Finding that perfect balance of ripeness – as ripe as possible but still capable of travelling from his farm to London and then onto our customers. Rather than imposing Natoora's needs, we changed the way we buy, giving him more time from order to harvest even though it means adding uncertainty to our forecast of how many tomatoes we need. This allows Raffaele to maximise quality and flavour. As simple as this may sound, it is a significant challenge that makes life more complicated for both Raffaele and Natoora. But it must be done if we are to protect the seed heritage we have built up over thousands of years. We need to modernise our approach within the context and realities of a global food system and build supply chains that are capable of carrying within them the heritage and innovation that brings flavour and quality back onto our plates.

Whether or not the Cuore di Sorrento is the best tomato in the world, what is important is the link between the genetic disposition of the seed and the characteristics that drive flavour. The selection that occurred over generations is at the heart of our agricultural practices, and the reason why seeds are, to me, the single most important contributor to flavour. Just like the quality of a pizza begins with the right ingredients and improves from there through technique, seasoning and baking, flavour starts with the right seed and can be improved on with the right soil and great farming practices. But if you start with highly refined and nutritionally deficient flour and use waterlogged tomatoes, you will never make a great pizza, no matter how good you are or how fancy your oven is.

As consumers, we need to understand the power of our daily choices, particularly the collective power we have. We can begin with identifying specific varieties. Learn to distinguish them, as they provide you with the greatest guarantee of joy irrespective of who grew them and how they were farmed. Collectively we can shift demand, which will see seed companies being tasked with developing and breeding flavour back into the system.

CHAPTER THREE

Protecting Our Soils

Cornwall greets us with glistening rays of sunshine. Still cool in the shade, it's the light that turns the monastic hedges upon which grow hundreds of wild plants into a glittering emblem of spring, the rays reflecting off the blades of grass, which backlight the wild primrose petals, while the white allium flowers of wild garlic and three-cornered leek hark back to wintery snow. Escaping the cold crevices of the stone walls, navelwort's water-logged fibres stare out at the sun yet sit in the shade of the nettles whose dark, musky deep green is broken by radiant dandelion flowers. The Cornish hedges are the most amazing edible monuments to spring, and it is the soil beneath that is responsible for the beauty above.

Flavour gives me more peace of mind than any label. It is not a guarantee of stellar farming practices or stellar soil health, but it is a strong indicator of the two. It tells us that the farmer is doing more right than wrong. Given the current realities of feeding an ever-growing population, mounting pest pressure, economic risk and sometimes hardship, which all need to be considered within the farming equation, I opt for flavour as the certification of choice. In the complex process involving hundreds of decisions over the course of a growing season, farming with a flavour-first approach leads to heighted nutrition and by association healthier soils.

Nutrition Under Our Feet

Having explored interconnectedness of flavour and nutrition, we must now delve deeper into the depths of the earth to the relationship between plants and soil. The importance of soil health cannot be underestimated. The correlation between flavour and soil health makes seeking out flavour a powerful weapon in the fight against the industrialisation of our food system.

We should all be asking ourselves how and on what do plants feed? This question is critical. What the vegetables in your soup consumed, the nutrients they absorbed, is the source of nutrition. Plants are living beings and they, like all living organisms, are engineered to seek nutrition.

What takes place under our feet is in the purest and simplest sense a collaboration between active soil biology – primarily fungi and bacteria – and the plant's root systems. The plant delivers goods and services from above ground in exchange for goods and services below. Fungi and bacteria want carbon while plants are after macro and, importantly, micronutrients, which fungi and bacteria provide. The relationship is far more complex, but this level of understanding is sufficient to appreciate two key points. The first is the dependency plants have on soil microbiology to absorb key nutrients; plants on their own are not capable of feeding themselves a wholesome diet. And the second is that a healthy soil is one that has a lively, thriving community of bacteria and fungi (soil microbiology) and most modern agricultural practices, like repeated heavy tilling, harm and deteriorate it. Once we understand the basics of this relationship between plants and soil we can make better food choices, as they will be informed by science rather than marketing. By following flavour, we can trace nutrition all the way back to the soil and ensure we are getting the maximum nutrition for our buck while doing right by the planet so our soils will be alive and rich for future generations.

When you consider that close to half of the earth's habitable land is used for agriculture, how we treat it is fundamental to the health of the planet. There is a direct connection between the planet's climate and the health of our soils. The last few years have seen a deepening of our knowledge of soil. We have a much better understanding of how it functions, how plants interact with it and the crucial role microbiology plays in this relationship. Interestingly, we have looked back in time to replicate what nature does if left to its own devices, and gained insights into improving soil health and therefore its ability to drive plant nutrition. A good example of this is mob grazing: a grazing system for cattle that mimics how pasture would be grazed by wild animals, which many ranchers are using. Similarly, we have a better understanding of what degrades soil and the negative implications its degradation has on our ability to produce healthy food as well as the impacts it can have on our climate.

The Case for Soil

Back in Cornwall, navelwort is poking its head through the crevices in the ancient stone walls that line much of the roads, while those vivid, pale-yellow wild primrose flowers are the strongest sign that the weather has turned and is now leading to summer. Sean, a wonderful friend and farmer whom we've worked with for over ten years, greets us at Liskeard station in short sleeves and tinted glasses.

A talented musician and poetic spirit, Sean approached farming from a somewhat unconventional path and perhaps this is why he was able to instinctively build a spiritual connection to plants and the soil they so depend on. After a career in the music industry as frontman of the Moonflowers and spending years in the countryside in northern France producing music, Sean stumbled onto Keveral Farm on the southern coast of Cornwall overlooking the village of Seaton. Here, Sean began his kinship with the earth on a piece of land known as Plot 7, growing baby leaves, edible flowers and young vegetables like baby beets and turnips as well as various varieties of radish, alongside lesser-known edible plant varieties like dittander, saltbush and pineapple sage.

Nowadays, the wider farming community is far more aware of the importance and benefits of healthy soil. Sean, however, is a different kind of farmer, more similar in style to Masanobu Fukuoka, the Japanese farmer and famed author of *The Natural Way of Farming*. A proponent of letting oneself be guided by nature, a practice of minimum intervention with nature's preferred pathways and inbuilt designs, Sean looks at the plant in its entirety, learns from it and understands what it needs. With no practical farming experience, he was free to focus on plant happiness and how his actions and decisions impacted the soil, and how the soil's health transferred to the plant's well-being.

Sean's intuitive, spiritual approach, as I understand it, created a profound commitment to soil health early on in his farming and has resulted in some of the most flavoursome produce I have ever experienced. In the early years, he focused on baby leaves, like mibuna and frill mustard, and microgreens – these are plants that are harvested while very young, when their size is tiny and the plants have just started developing their first leaves.

Microgreens became a thing in high-end restaurant kitchens in the mid 1990s and are still used today. The catch is that most microgreens are not very tasty, lacking any length of flavour, and the issue is they are not grown in soil. All commercially available microgreens are grown in a substrate. Recycled textile fibre is a common one. Sean began experimenting with different mixtures of compost, always focusing on plant health and flavour to guide his experimentation. The flavour he achieves is incomparable to what is readily available, far stronger and complex, and the shelf life is twice as long. Both traits are a result of the soil's impact on plant growth. While compost is technically not soil, good compost, like healthy soil, is teeming with biology. In contrast, textile fibre is there as an anchor point, providing no real nutrition to the plant, and growth is almost exclusively down to water with added nutrients. These plants grow faster, evident from the length of the stems at such a young age, their flavour is terribly lacking and waterlogging means their shelf life is limited. Once cut from the tray, they have little substance to survive.

I vividly remember the day Sean and I met for the first time in his home on the farm. At around one o'clock in the morning we were still deep in conversation when Sean paused. 'You must try this spinach. My friend Brian grows it not far from here. He still farms with horses.' The Amish in Pennsylvania still practice this kind of farming, but few outside these communities do. Farming with horses is low impact compared to tractors – they are gentler on the soil and therefore protect the biology beneath.

Off he went into the dark of night and came back with a bag tightly wrapped around his fist. Out of the bag came a vibrant, meaty spinach, which he cooked with a knob of butter and salt. Within minutes, we had a plate of beautifully simple spinach – to this day, the finest spinach I've ever had. The deep minerality and complex vegetal notes, balanced with the fantastic, earthy sweetness that is characteristic of great spinach could not be explained by the plant's genetics alone. This was the work of nutrient-rich soil with active microbiology that enabled the plant to absorb a wealth of minerals and nutrients that would best express its already strong genetic profile.

Sean and I walk through the fields often, in the sunshine and in the pouring rain, all beautiful in their own way. If I close my eyes and think back to those days in early spring, I can feel my waterproofs stroke the wet, strong kale and collard plants, string instruments creating a harmonious squeaking rhythm. Boots to the ground, pressing a giving land, the pounding of wet earth forms a structural percussion sound, and the coastal fronts deliver a grand wind section. In an instant, an improvised symphony is playing all around us, in harmony with nature – the pleasures of creating beautiful food. There is symmetry in a well-cared-for plot of land and it speaks to the life below. The health above ground reflects what is going on underneath. Complexity is a word I use regularly to describe flavour. Complexity of flavour, tasting more than sweet and salty, or acidic, occurs because what we are eating contains a wider mix of macro and micronutrients, amino acids, vitamins and so on. To achieve flavour complexity, in other words, flavour that makes you go 'wow', a farmer must be working with healthy, living soil. When you bite into the sprouting top of a broccoli plant or the stalk of a collard leaf in Sean's field, you feel good. It is your body telling you that this is nutritious.

Understanding Soil

Soil is essentially ground-up rock. Sitting on the surface of the earth, it is made up of water, air, minerals and organic matter. Organic matter represents the smallest percentage of the four and can be anywhere from 1 per cent to as high as 40 per cent, but it usually sits in the range of 5 to 8 per cent. Soil types vary based on the type of ground-up rock they are composed of and are heavily influenced by geographic location. They are classified according to the proportions of sand, silt and clay they contain. Water runs through sandy soils with far greater ease than soils rich in clay, which are great at water retention. Living in the soil are billions of organisms that help define the structure and properties of the soil; in turn, these are further modified by human impact on the land.

Farming practices have the greatest and most severe impact on land quality and its properties. A great example of the role soil plays in mitigating severe weather events is its ability to absorb and retain water. Flooding and drought are the two most common weather-related problems farmers face, and they can be devastating, wiping out entire crops, damaging infrastructure and damaging the soils themselves. A soil's ability to avoid flooding and, perhaps counterintuitively, drought is down to its structure. Think of it as a sponge: the more air pockets contained within the mass, the greater its ability to both absorb water into it as well as retain it once it's there. Soil microbiology plays a fundamental role in breaking down soil from a compacted mass into a cakelike structure that is super resilient in the face of adverse weather. Industrial agriculture uses heavy tilling as a key process in field preparation. By ploughing a field, you release precious carbon into the atmosphere as it makes contact with oxygen, you disturb the soil microbiology, you accelerate the breakdown of organic matter and the heavy tractor moving through the field compacts the soil.

From the perspective of human health, we know that soils that are teeming with life are far better at delivering nutrients to plants, including micronutrients, which are fundamental to our health. Elements like zinc and iron are lacking from diets worldwide and this problem faces developed nations as much as developing ones.

If we are to ensure a healthy future for our planet and for ourselves, don't bother with labels. You must seek out flavour as a strong indicator of a commitment to healthy soils and 'better than most' farming practices.

The World's Sweetest Onion

It is mid-May, roughly thirty days from harvest of one of the most spectacular onions I've ever come across, with a natural ability to grow to over one kilogram in weight. The Breme onion's physical simplicity is perhaps its cloak, for underneath those rugged, atypical outer layers lies the sweetest onion in the world. There is possibly a sweeter onion out there, but I have yet to come across it. The flatlands here absorb the intense, humidity-drenched heat, so very quickly we find ourselves huddled in the shade of a large tree, sweating as we stand.

The town of Breme has a population of roughly nine hundred inhabitants and is in the province of Pavia on the southern part of Lombardy in northern Italy, almost on the border of Piedmont. Many centuries ago, the Sesia river was deviated to make way for fertile agricultural land, a common practice back in the day. Cultivation of Breme onions takes place on what was once a riverbed, exclusively in an area four kilometres long by five hundred metres wide. Fascinatingly, if the onions are planted on the fields that lie on higher ground in Breme, the flavour is unrecognisable to those grown in that very specific area. Logically, this speaks to the quality and very specific characteristics of the soil in these two square kilometres of land.

This tiny area of land is tremendously fertile – as an old riverbed, its fertility is derived from the soil's composition and therefore its history. Rivers carry water downstream into seas and oceans, but they also carry sediment, rocks, soil particles, organic matter and nutrients. Over time, there is an accumulation of nutrients along the ground and banks of the river as all these deposits settle into the earth.

The history of the onion's cultivation goes back to the tenth century when it is said the Benedictine monks started growing them. One thousand years of seed selection and adaptation to this specific soil. This is extremely important. The interaction between the onion plant and the soil biology has been fine-tuned over centuries, creating an almost perfect relationship. The vast nutrient deposits are what give the otherworldly sweetness to the onion.

Young, passionate and committed, Luca is toiling under the dense sunshine with a small home-garden rotavator, weeding between rows of purple onions. Hardly a life of luxury, Luca's is a life of love, a life full of meaning. In his grass- and dust-covered smile you can feel his joy. It has been a pleasure working with Luca all these years after discovering this little-known onion. The cultivation methods, which are mostly unchanged since the time of the monks, have a beneficial impact on soil health. Luca is adamant that these techniques are crucial in sustaining the soil's fertility, which itself is the key to the Breme onion's remarkable taste. Maintaining the diverse nutrient structure, and in particular the high levels of potassium and phosphorus present in the soil, is front of mind for Luca. Here is an example where the soil is as important as the seed.

These special onions are cultivated exclusively by hand, except for the weeding described above. This low-impact approach to farming, with no chemicals, minimal inputs and gentle soil preparation, are fundamental to fostering soil biology. Cover crops like radish are used to build organic matter and reintegrate essential compounds and are then mulched and worked back into the soil. Manure is then applied for added fertility and is also worked into the topsoil. Luca's tilling is thoughtfully shallow for minimal disturbance to soil life as he prepares the ground for cultivation.

The connection between the sweetness of the Breme onion and the thriving soil biology is evident. To understand this, we need to look at why some onions are sweeter while others are spicier and more pungent. Onion genetics are primarily responsible for flavour and therefore sweetness; however, it is the amount of sulphur compounds – made from sulphur absorbed from the soil – in the onion that impacts sweetness. In simple terms, the lower the sulphur content, the sweeter the onion.

Sulphur is a common chemical used in onion farming around the world as it produces an enzyme that is responsible for chlorophyll formation, which stimulates photosynthesis. It also increases the plant's ability to utilise nitrogen and is seen as key to driving yields. Luca entirely removes the need to add sulphur by protecting the quality of his soil, thereby leveraging the richness of his land. Not only does he have the Breme's genetics on his side, but the highly fertile soil also reduces the sulphur uptake of the plant, further enhancing its sweetness. I find this complicity between nature and farmer, the interconnectedness among seed, soil and farming ethos super fascinating. It is like a beautiful tango; sadness and humour weaving together to create a work of art. Flavour is the product of a similar artistic endeavour.

Locally, it's known as Cipolla Rossa di Breme (red onion from Breme); however, I would have named it the pink onion, for its skin is a delicate shade of profound pink. Even after a good hour of cooking, this pink remains ever present, albeit subdued and caramelised. Bremes are exceptionally crunchy when eaten raw, their shape is squashed and fat, their size is massive and contradictory. The soil's high potassium and phosphorous content is responsible for these characteristics. You'd never think a vegetable this oversized could be so incredibly sweet and elegant; this is possibly the same reason the locals swear by its high digestibility. Unlike its cousin the red onion from Tropea, another famed sweet variety from Calabria in southern Italy, each layer is significantly thicker. Its layers are so thick that at first sight they seem overgrown, as if the onion had been forced to grow through excessive water and fertiliser applications. To the untrained eye they appear spongy, and you might expect them to be tasteless. Interestingly, these physical attributes, what almost look like malformations, are its prized characteristics. They are signs of its supreme quality that can only be attained by growing them in this unique piece of land. Soil fertility and the right seed genetics are irreplicable.

The Breme onion is a pristine metaphor for the art of sourcing flavour-first produce. A thousand-year-old seed, adapted to conditions so specific as to exist purely in a region the size of several football pitches, grown by a handful of people on Earth within a completely closed circle of life almost exclusively by hand, purely for the joy of it; to retain and prolong tradition, culture and insane flavour.

Yet it does not end here; this onion exemplifies why sourcing and eating these varieties is so enriching. It educates us through its appearance that belies an insurmountable flavour, reminding us to be open and accepting that every single plant variety and part of a plant can excel and be utterly special. And therefore, this humble onion is a metaphor for hope. Hope in the ability of humankind to fix this broken farming cycle that is so damaging to our health and that of our planet.

A View into Industrial Organic

Apples are perhaps the finest example of the distortion that has taken place in the organic sector and, crucially, the value of the organic label to the consumer. How much can we rely on organic certifications? It is worth exploring as it clearly exposes the vast gap between the original vision of the organic movement, which is what consumers have in their mind, and the reality of how most organic produce is grown today.

My friend Magnus, who was the chef at the world-renowned Fäviken restaurant in Sweden, is now living with his family on a stunning, super diverse apple orchard in southern Sweden. The restaurant was located on the Fäviken Egendom estate in a remote part of the country. That seclusion and immersion in nature meant that Magnus relied heavily on ingredients that were foraged, grown or caught on site, while working closely with farmers and fishermen from nearby. Magnus and his team built a close connection to the soil and an understanding of the ecosystem. Of course, it was inescapable that he would find harmony with nature in this setting. I believe this experience strengthened and evolved his thoughts on natural farming methods, which would later be instrumental to decisions he would take in the orchard.

The vision of organic agriculture, which dates to the beginning of the twentieth century, has been brutally bastardised by the modern food system. Sabotaged, infiltrated and destroyed from within, it has left consumers with a useless and ineffective tool to guide their purchasing. It has also left 'real' organic farmers with a watered-down system that is unfair from a competitive standpoint and betrays the principles they stood for when they first created the system.

From any angle you care to look at it, an organic certification today is unfortunately not a guarantee of good farming practices. It does not tell the consumer that the farming of that product was beneficial to the soil it grew on, nor can it be relied on from a nutritional density perspective and, sadly, the same is true of flavour. Furthermore, ultra-processed foods, which are the source of most major health issues today, can easily be certified as organic. Consumers are being lied to.

When Magnus first took on the orchard a few years ago, his intention was to farm apples in what he describes as a traditional industrial organic way. One hundred per cent of the organic apples that end up on supermarket shelves have been heavily sprayed with copper or sulphur, or both, to combat a fungal disease called apple scab. They are chemical elements that are extremely toxic to soil microorganisms when they go out of balance. Additionally, evidence has shown that copper can have harmful consequences on human health. This is precisely the opposite of what consumers believe they are buying when they pick up a bag of organic apples.

The stringent cosmetic expectations demanded by supermarkets are the culprit because apple scab, a fungal disease that affects apple trees around the world, causes physical scarring on the fruit, which is unacceptable to them even though the fruit is perfectly safe to eat. Copper and sulphur are used extensively to combat apple scab in organic orchards as they are approved fungicides. In Sweden, where copper is not permitted, farmers use sulphur and sodium bicarbonate, but in most countries, copper is still widely used. Copper is not only a heavy metal with adverse consequences on soil health and plant vigour but its low solubility means copper toxicity is hard to correct. Ideally, we don't want our soils drenched in copper or sulphur and yet industrial organic agriculture is totally dependent on them.

As Magnus dug deeper into these practices and what that meant for the soil on his farm, he felt the need to find another way. On top of a decline in soil health, a traditional commercial crop of organic apples would have led to less diversity of plants on the ground around his trees, fewer insects and pollinators, an ecosystem in decline rather than a thriving one.

Magnus chose instead a path of no spray. It helped that his background in the kitchen and fermentation opened the cider market as a real commercial possibility. Not only selling apples to cider makers but farming beautiful, diverse fruit to produce his own cider. Like Sean, the innate desire to improve soil health to drive flavour led Magnus to discover how nature in a thriving ecosystem can tackle apple scab with zero chemical intervention. It required him to take a step back and observe nature doing its thing.

Nature always proves smarter, and Magnus realised this by looking at the full cycle of life in his orchard across the whole year. His decision to ditch the commercial organic apple dominated by supermarket cosmetics for cider, where fruit cosmetics are irrelevant, allowed his orchard biology to thrive. As the applications stopped, insects, bugs, plants, worms, the whole biology of the orchard came to life and grew in numbers and diversity.

This new activity all around the orchard on the surface of the soil was a game changer and spoke to the activity below ground as well. There is a whole life cycle that occurs at that level in harmony with the seasons. As trees fruit and the weather gets colder, leaves begin to fall and decompose. Different insects appear and disappear, bees stop working and hibernate, and fungi go dormant, awaiting spring to reawaken. Apple scab present on the tree's leaves and fruit lives on, staying dormant as they decompose at the foot of the tree. In the spring, any leaves that have not fully decomposed will be an immediate source of apple scab and, since the decomposing of the leaves is the work of the biology within the orchard, all those copper and sulphur applications stunt the ecosystem's life and its ability to fully break down the apple scab-infected leaves. What Magnus started seeing on his farm was the complete decomposition of apple leaves by the time spring came around and, as if by magic, his incidence of apple scab reduced. The attention on soil health was fundamental. As is Magnus's persistence with flavour through diversity, best-in-class farming practices, and simply because he wants to make fantastic cider. The connection between flavour and soil health is indisputable.

Labels have become meaningless. Even worse, they trick and fool you. What really counts is the spirit of the farmer. Demanding flavour is a salute to soil health and the health of our planet in a way that no label can achieve. While we've mastered the art of formulating artificial flavours for processed and ultra-processed foods, we should not be adulterating fruits and vegetables, meat, dairy or any raw ingredient.

If we want a healthy planet, we must treat the land with respect. At 44 per cent of global land use, agriculture must move towards a soil-first approach. How we grow our food has an overwhelming impact on our soils and climate, with the global food system accounting for roughly a third of the world's greenhouse gas emissions.

As a consumer it is impossible to see for yourself if the apple you are about to buy has been farmed with healthy soil in mind, yet thanks to the interconnectedness of flavour and soil health, flavour is our key to demand an apple where soil health is actively considered and acted on. By seeking out joy as a way to satisfy ourselves, nourish ourselves and do well by the planet and future generations, our palates will evolve. That evolution, that re-education is the way to regain our collective independence.

CHAPTER FOUR

Character and Flavour

If flavour is the key to the food system revolution, then uncovering flavour visually is equally important. Since we cannot take a bite out of every pear to judge which one tastes better, we must learn to use visual cues to select the fruits and vegetables we consume. We have lost this skill in tandem with the decline in cooking; if we are to regain our ability to really feed ourselves, we need to start with an ability to select the ingredients we want to cook. All raw ingredients tell you something about how they taste through their physical appearance and, by virtue of that, how nutritious they are as well as giving you an indication of whether they've been farmed intensively or by someone who cares. In most cultures, we were taught to buy food by our parents and grandparents in the same way cooking was passed on through generations. Seeing how they selected a bunch of chard or which three lemons they chose from a box, and which they rejected, was an integral part of our culinary education. I learned how to select apples from my father: by listening to its sound as you flick it, it tells you whether its texture is crisp or floury.

I find the parallels between plants and humans to be truly fascinating; we are all a product of mother nature and there remains a remarkable similarity. It is in those terms that I want to explain a way of looking at food that can have a lasting impact on our food choices, as well as effect the wider shift in consumption that is needed. Nothing less than a new way of looking at the food we eat.

We need to retrain our eyes to see beauty in what we've been told for decades is ugly. Seeing flavour in the physical 'imperfections' as natural perfection and the beauty and diversity they express. The same physical diversity that defines the human population is true, in some respects, of the fruits and vegetables we eat. The concept of uniformity that is so prevalent in our food culture today is wrong and responsible in large part for the dilution of flavour we experience today.

Ditching Uniformity

Imagine a world where we all looked the same. It would be alien to us and horrifying. Why does a species that seeks diversity in everything, and used diversity as a survival strength, now seek homogenous plastic-looking food? The answer is less important than realising it is counternatural, harmful and debilitating to our intelligence. What makes life incredible is diversity and contrast. The richness of our cultures is why being human is so beautiful because it fills us with spirituality and meaning, it broadens our minds and makes life interesting. Few of us would give that up, yet we have given it up in our foods. Nature has symmetry and a degree of uniformity built into it – nothing evokes this more than fractal symmetry, which we see in a cauliflower, particularly a Romanesco, or a snowflake – but we know that nature thrives on diversity.

When humans began cultivating the land, we domesticated plants and animals and, in doing so, we reduced the number of species we would reproduce. We also vastly increased the quantity of food by planting rows and rows of plants from that reduced subset. Therefore, the development of agriculture was the crafting of a system that led to a degree of uniformity; however, modern agriculture has taken uniformity to a completely new and unhealthy level. Those rows of identically shaped and coloured peppers are not what nature intended us to eat, and are certainly not beautiful. Beautiful is edgy, scarred, mysterious, weird, different.

We can thank our global supermarket culture for the uniformity in our food lives. What they demanded from farmers pushed seed companies to breed in yield, uniformity and this new concept of beauty and, in turn, breed out flavour and nutrition. In their book on farming practices and soil health, David R. Montgomery and Anne Biklé contrast this practice against the reduction of minerals in the soil itself: 'Far less controversial is the effect of crop breeding in trading off nutrient content for high yields. Though it is not widely acknowledged beyond agricultural circles, this well-known reality has a name among agronomists and soil scientists – the dilution effect.'[1]

Supermarkets introduced a further filter at the warehouse gate: exacting standards of what is and isn't acceptable. As we saw with organic apples, physical standards have consequences in how we farm. Simplistically, that is how we have ended up with bland, uniform, alien-looking fruits and vegetables in our lives and, I should add, *ugly*. There is a company called Imperfect Foods (previously known as Imperfect Produce) whose purpose is to sell produce rejected from supermarkets. They should have named the company Perfect Foods instead, as that would push consumers to reconsider what is wrong.

So here we have our first parallel with humans: perfection is diversity. When you see rows of identical apples, you can be certain they will not be the tastiest. Those identical peppers that look like they came off a factory production line will not be the most nutritious or delicious.

The second parallel with humans is that character is richness. Richness can be felt through flavour and nutritional content. If you were to pick an individual to spend an evening in conversation with, my guess is that most of us would pick the bearded, scarred sailor with a hook for a hand, rusty wind-blistered skin and a twinkle in their eye at the end of the bar. Those physical attributes are a sign of a life lived to the fullest, a life full of stories, a life that is interesting and rich. These are exactly what we need to look for not just in produce but in most of the food we eat. If you are to pick an apple, chose one that is a little off-looking, that stands out for its peculiarities. Seek out character.

Some of these physical attributes in a fruit and its increased sugar content can be scientifically explained; however, not all links between what we can define as a character trait in a vegetable and flavour can be answered with science. I like to believe there is a metaphysical connection, a way of nature, that just *is*. Asparagus scarring is an interesting example. A little-known secret is that curved asparagus spears, and particularly those few that have scarred when they bent, are the tastiest ones. Spears will usually bend due to the wind, and not all spears in a field are exposed to the same force of the wind. Those bent spears that bore the brunt of it will sometimes scar on the side where it curved, opposite where the wind was coming from. The asparagus that had to fight to survive is the tastiest one, has more nutrition and has the scars to prove it. There is a plum in France whose scarring is emblematic of its variety, and behind that lies the answer to their mesmerising sweetness.

The Scarred Reine Claude Dorée

As we drive up the narrow winding road, dust rises and settles, billowed by an incessantly dry current of air that navigates across the plains and up through the hills, raking leaves and heat-scorched petals across a landscape that's idyllic in its free spirit. Federico and I have always been moved by the plums that grow here and we've been working with Marc for several seasons, but this is the first time we are here at the tail end of an extremely short Reine Claude season. We are filled with anticipation.

A consummate farmer, Marc's skin bears the rough, matt surface that provides a stark contrast to his glistening, sparkling eyes. Great farmers, you'll notice, speak to you with their eyes. His body is strong, weathered and half of it is on show as the sun is blistering. His hair is a million shades of grey, but there's plenty of youth in Marc; you can sense it in his movements, his pace and agility while the tone of his voice shows no signs of wariness.

The land is made up in large part of clay mixed with rock, most of it calcareous. Like many of the finest vineyards, these soils are poor in organic topsoil and rich in minerals at the same time. They are great at producing quality fruit. The rock content and poorer fertility forces the plants to go deeper and work harder to find nutrients, but that same composition gives the fruit these wonderful complex mineral notes. What we see here is typical plant stress caused by the natural environment. The plant's nutrient needs forces it to dig deeper into the soil to find sustenance, and it is widely known that this heightens flavour and nutrition. We can liken this to living a harder life – trees that work harder to survive produce better-tasting fruits.

Reine Claudes are a subvariety of plum grouped under the name greengage in the UK and the US. Some believe that all greengage varieties can trace their origins back to the Dorée. It is important to distinguish that there are a few varieties of Reine Claudes, which are different to the Dorée or 'golden' variety that Marc grows.

As we've seen, varieties are fundamental to flavour, and this is a case in point. Dorées are genetically predisposed to scarring at the end of the ripening process because of the fragility of their skins. As the sugar content in the fruit reaches extremely high levels the skin splits, like a cut on your finger. It is along these cuts that scarring occurs, in the same way as our own skin repairs itself. More modern varieties of greengage are thicker-skinned and therefore rarely, if ever, scar. Also, their ability to convert nutrients into sugar is not as good so they are not capable of reaching the same crazy sugar notes of the Dorée. When fully ripe a Dorée is on another level, not only compared to other greengage varieties, but to most other plum varieties in the world.

Genetics on their own, and the quality of the soil, are not enough to achieve fruit that is so exceptional. Marc's nerves of steel and commitment to flavour push him to keep the fruit on the tree for longer than most, at significant financial risk. The longer the fruit stays on the tree the more delicate it becomes, and heavy rainfall during the last stages of ripening can be a disaster, as the rain that accumulates in the dimple where the stalk attaches to the fruit can seep into the fruit and cause it to burst open. The reward is worth the risk for a farmer who is passionate about putting quality above the financial risk. We should not overlook his commitment.

Scarring on the plums occurs in a circular pattern at the top of the fruit, around the dimple. Concentric circles appear as the scars form: they leave thin, greyish-brown lines around the stalk. Sometimes, different-shaped scars are visible on the plum, but the location and look of these are the most iconic. Very similar to the scarring that occurs on older tomato varieties, these scars add a layer of beauty to an already stunning-looking fruit. This specific plum variety is a wonderful example of what could be lost through the pursuit of uniformity, and its physical appearance is a testament to the need to retrain our eyes so we can develop a new way of looking at the food we eat.

As we wander through the orchard, the colour of the plums at first escapes us, as everything around is so green. The canopy of leaves, the grass and weeds underfoot, the mossy bark and fresh vegetation: all varying shades of green. As we walk down the row of trees, it is hard to discern the fruit from all the vegetation. That is until you see the first speck of red. Once your eyes catch that first fruit they readjust, refocus, and suddenly more fruit appears as your senses learn to distinguish the small plums that hang from the branches like ornaments. The outer edges at the top of the tree are where you find the ripest fruit, where the sun exposure is greatest. The juxtaposition of cold nights and warm days stimulates the production of anthocyanins, pigments that are responsible for painting the reds, blues, purples and blacks on the fruit's skin.

Taking one in my hand this late in the season, the full characteristics of the fruit become apparent. Scarring along the top in the form of thin circular lines that surround the stalk with a rainbow of colours from the deepest purple to the lushest orange, with plenty of green in between. Inside you will find flesh with an intense yellow-orange glow that defies reality.

Other greengages simply don't behave in this way; they have been bred for efficiency at the expense of flavour whereas the Dorées, well, they've been built to leave you breathless even before the first bite. Their character and physical expression are a sign of their wild side, their desire to take a beating and tell the story, a story that is evoked through an intensely complex flavour in comparison to the one-dimensional flavours of a timid, reclusive variety like the common greengage. The risk Marc takes to leave the fruit on the tree for longer pays off because the genetics are there to turn a very good plum into an extraordinary one – those last few weeks are game-changing. Bite into one of Marc's wild-tempered plums and they will transform you.

The blemishes, the signs of stress, skin colour and tone, the scarring – all these physical signs tell a story of how the plum has grown and its individuality. The more such clues, the more character, the more flavour, the more nutrition.

A fruit of immense character and physical beauty, a humble plum can teach us important lessons on how we look at the food we eat, and what those choices mean for the future of the food system, of our own joy and pleasure, and the planet we will leave behind for future generations. We must look at food in a fundamentally new way.

How Stress Develops Character

Among artichokes and tomatoes, a scattered lemon tree and the most beautiful hibiscus flowers on this planet, an undiscovered legend is producing some of the most extraordinary, flavourful food in the world. Romualdo is a true artist and master grower with a 360-degree understanding of nature. His approach to his craft is so unique, so complete, it is enlightening to walk through the fields with him.

Romualdo is a gentle man, whose warm heart shines through his voice, and in his hands he stores much of his knowledge, where it feels safer. Romualdo has really understood the Sardinian technique of growing tomatoes in the winter, exalting the characteristics of umami-rich flavour, dense firm flesh, crunchy skin, sweet and acidic flavour balance, and contrasting colour that beautifies and catches the eye.

The farming of tomatoes in winter is perhaps the modern growing technique displaying the mightiest innovation in flavour. Character is at the heart of winter tomatoes, and plant stress is how character and, in turn, flavour is developed. If you do a search for 'plant stress in farming', the results will discuss the link between stress and negative impacts on yield. Temperature extremes, lack of sufficient water and, as we will see as we delve deeper into Romualdo's techniques, the salinity of water are all cited as causes of decreased yields. Plant stress within traditional industrial agriculture is seen as a problem to be avoided.

By contrast, many of the farmers we work with actively seek to stress the plant precisely to impart flavour through hardship. As we see in ourselves, hardship and overcoming difficult situations builds character. Intentional plant stress has been practised by farmers for generations as a technique for improving flavour and productivity, utilised widely in Mediterranean growing regions. This practice is most revered and extensive in tomato farming and, in my view, some of the most innovative farming techniques of this century are the stressing of tomato plants from an obsession for flavour. In chapter 7, I share the story of this pursuit and the patched history I have managed to piece together over many years due to my own obsession with tomatoes. What is certain is that conditions that test the plant's resilience and will to live, just like us, result in stronger, more complex character.

Plant stress is more scientific and encompasses a wide variety of external factors that affect the plant and cause internal responses that usually result in heightened flavour. In many instances, stress will manifest itself in physical form by altering the colour or shape of the product. In the asparagus example I described earlier, the plant suffered a form of stress – abiotic stress to be precise – from the wind. Plants can suffer from two forms of stress: abiotic, which comes from environmental factors such as light, water salinity, heavy metals and temperature; and biotic, which comes from harmful organisms such as insects, pests and fungi to name a few. One of the most widely known stress-related facts is that brassicas (kale, broccoli, cabbage) sweeten when exposed to cold weather. The stress caused by below-freezing temperatures forces the plant to convert starch into sugars, as sugar lowers the freezing point (this is the reason ice cream exists) and therefore the kale plant's cells can survive the harsh cold. This concentration of sugars is present throughout the plant but is most noticeable in the leaves and stalk. Nature likes to reward the will to survive!

The most world-renowned areas for growing tomatoes are by the coast, close to the sea. One of the benefits is the natural salinity of the ground water, which will undoubtedly be significantly higher than in a mountainous region, for example. The salinity in the water, which is measured via electric conductivity, can range from 250 µS/cm (microsiemens per centimetre) or 165 ppm (parts per million) in normal conditions and reach up to 5,000 µS/cm or 3,200 ppm near the coast. Salinity does two things. Firstly, it adds complexity to flavour – think of it like pre-salting your tomatoes. Secondly, it stresses the plant. Most of the plant varieties in cultivation by humans cannot handle elevated amounts of salt as it interferes with root osmosis – the water being too dense for the roots to absorb it. Winter tomatoes rely on high-salinity water to impart the much-desired stress; however, this process needs careful management.

Romualdo has figured out a way to stress his tomato plants as far as possible in pursuit of flavour. He takes minute details into consideration, ranging from soil composition and ambient humidity to light intensity, and how these impact plant stress. For example, he has noticed that on brighter days he can feed the plants water that is saltier. He told me he is not entirely sure why and is still trying to figure it out, but he does notice how the plants can handle the added stress on brighter days.

The level of detail in Romualdo's techniques is fascinating and worth exploring to understand the commitment needed to develop character in the fruit. Romualdo has discovered what he calls an addiction to salinity, where plants begin to suffer if the salinity is not high enough and therefore need weaning off gently to reduce excessive stress. While stress is the desired outcome, there is a level of stress that is counterproductive. On rainy days, due to increased humidity in the air, the tomato plants absorb this water through their leaves on top of what their roots are tapping into in the ground. This additional 'sweet water' from the ambient humidity causes an imbalance in salinity and can stress plants more than desired, so Romualdo has learned to monitor these delicate balances daily. Romualdo collects his own rainwater and taps into a well, which, being close to the sea, has water with high salinity. These two water sources allow him to create a specific blend each day depending on his plant's needs. Every day he decides exactly how salty the water he feeds his plants needs to be, the level of stress he seeks to impart and the changing environment.

When it comes to irrigation, he is just as considerate. He knows he can water at different times of day and that he can vary the quantity, but he can also decide where he waters them. Commonly, irrigation occurs at the base of the plant via an irrigation system that feeds water directly into the root system. Romualdo irrigates inches away from the base of the plant, forcing the roots to search for the water to stimulate root growth. Roots grow longer, stronger and healthier. This is pure genius as it enables him to elevate the levels of stress the plant can take; the stronger the plant, the more adverse conditions it can handle. All in the name of flavour.

It is time that we all learn to look at fruits and vegetables in a different way, to take the time to fully understand them and their peculiarities. By slowing down and engaging with our food we can in turn feed ourselves and our families healthier, more satisfying food. A coastline is more expressive if you take time to look at the detail within it, and that requires you to slow down and focus; if you speed past, you miss all the gorgeous subtleties. The same is true of raw ingredients, especially fresh ones – you must take your time.

The Role of Water

Water is life. When used judiciously and with intent to maximise the quality of the food we produce, water is an insanely magical resource. The water cycle is a highly complex system and is responsible for so much of our raw experiences – if you've ever had the thrill of being in a tropical thunderstorm you can feel nature at its grandest. As you see and feel unimaginable quantities of water falling from the sky, you can visualise what plants in these lush ecosystems experience and understand the richness of life. It is no coincidence that in the areas of this earth where sun and water are at their mightiest, the vegetation is so overpowering and so enchanting.

The flipside of this beautiful source of life is that if I had to summarise the single greatest impact agricultural industrialisation has had on the quality of fruits and vegetables, it would be the loss of flavour, which as we know equals a dilution of nutritional density. And if I were pressed to identify the main culprit for the decline in flavour, I would say it was down to excessive watering. This statement could be viewed as a gross simplification, and in many ways it is, but it does help to explain, in simple terms, what I believe to be the biggest culprit in the dilution of our produce.

Of course, excess water is not the root cause, it is a symptom of the system. But the reason most of our produce is tasteless is because of our obsession with yield, an obsession with producing enough food to feed the world. We have done this with total disregard to the quality of what we produce and the ramifications on our health and the environment. I have done my fair share of tasting products side by side, and the first description I'd use to describe the flavour profile of standard supermarket produce today is 'watery'.

It is important to learn to distinguish the hallmarks of watery fruits and vegetables, produce that has not developed character. Supermarkets are good places to study this as most of the fruits and vegetables sold there are varieties where character has been bred out, but they are good at absorbing water and fattening up quickly. Spend some time looking at the colour of the fruit, especially how pale or vibrant colours are on the outside as well as on the inside when you cut into it. When you sauté a courgette, pay attention to how much water is released in the pan. At Natoora, we source incredible cave-grown mushrooms in France that take time to mature. They barely release any water and cooked with just butter and salt taste incredible. Next time you cook some button mushrooms from the supermarket, look at how much water is released in the pan – you paid for that weight and the lack of flavour is a sign of nutritionally deficient food.

Pick a few tomatoes and taste them side by side to see how certain traits can guide you to better flavour. Our quest for yield has left us varieties that are superstars at growing fast but are dumb as hell. The same happened with chickens thanks to the Chicken of Tomorrow competition that ran as a series of contests in the US from 1946 to 1948. It was set up by the United States Department of Agriculture and supermarket chain A&P, with support from the major egg and poultry players in the country, to engineer a chicken capable of growing bigger and faster – as they called it, a 'better chicken'. Layer on the market dynamics that incentivise farms to produce tonnage versus quality, and it is no surprise that produce tastes watery.

While studies have not found that this dramatic increase in water content is directly related to the decline in micronutrients, water content is rarely examined and, given the naturally high levels of water in vegetables and fruits, small changes can have meaningful impact. Water is the go-to tool when applying stress to a plant – reduce water beyond a point and the plant begins to stress and develop character. When you try fruit from a tree that has not been irrigated (usually found on the edge of a field) you can taste and feel the concentration of flavour. The difference in texture is tangible, the lack of water evident.

The importance of water cannot be overstated. Water is a natural diluter of flavour; it gets in the way of complexity and removes nuance. Overwatering is a common tool in today's industrially farmed produce, particularly in areas where water is cheap, as it is an effective tool to increase yields per acre. Where commercial tomato growers in Florida are watering plants twice a day, the growers we work with in Sicily are watering anywhere from once a week to once every two days, depending on the season and the plants' needs. What you as a consumer end up buying is a lot of water and very little nutrition – a great scam. We then fortify foods through processing to compensate for treating food as we treat the manufacturing of hard goods, with an efficiency and uniformity mindset.

In the field, you can compare, side by side, over- and under-watered plants; of course, you can taste the difference as well. But, purely on visual inspection, you can see the same differences I talked about earlier. In tomatoes, as with most fruit, the physical cues are easier to spot, more evident to the eye. What you'll see is larger fruit size, paler colouration and increased translucency due to its watery flesh, plus less ridging if the variety has that genetic and, overall, less character. The better-grown plant, let's call it, will be dark, deep, rich, ridged, scarred, small – all the characteristics that you're looking for, while the plant that had a broken irrigation pipe and received way too much water will look like a supermarket tomato – light, pale, anaemic, round and bloated.

Water, and the flavour dilution that comes with it, is one of the issues in vertical and hydroponic cultivation of our food. We cannot replace soil functions, nor can we mimic a highly complex microbiology whose operation we have only just started to comprehend and likely will never fully grasp. These artificial growing environments have some benefits and likely can fill gaps in the food supply chain, but I find the same issues with flavour dilution due to excessive water and lack of soil micronutrients that are key to providing complexity in how we experience food.

Furthermore, there is a solution to both the climate crisis and the agricultural crisis – which, remember, produces up to a third of worldwide greenhouse gases – but it sits beneath our feet. By utilising the soil, working with it and nourishing it, we can farm in a way that is beneficial and constructive to the planet. Rebuilding soil fertility, drawing carbon down from the atmosphere, improving the water cycle, building climate resilience, increasing nutrition density in our food – these are powerful incentives to farm as nature intended, in harmony with the earth's soils, rather than finding ways to farm in recycled shipping containers with LED lights.

Spring Water and Watercress

Watercress is a plant that was born to live within water and the best growers give us an interesting insight into water management and building flavour. Ed is a man who belongs by the sea, sitting atop his boat or with water beneath his toes. In West Ashling, near Chichester, Ed's family have had their boots ankle-deep in spring water since 1835, growing watercress at Hairspring for generations – making them one of the oldest English growers I've worked with, if not one of the oldest anywhere in the world.

More so than any other produce I've come across, Ed's watercress fully embodies the concept of terroir. Hairspring sits pretty in the Chalke Valley, and even the sound of the word 'chalk' allows my mind to connect to the rich minerality of Ed's watercress. Its unassuming stalks and leaves hold a weighty heritage and its flavour is infused with almost two hundred years of farming tradition. This watercress is bound to this natural spring, in this specific corner of West Sussex; there's no prying the two apart.

From the car, I already catch a glimpse of the ancient stone beds, purpose-built for the farming of watercress hundreds of years ago. A patchwork of flat stone squares is sunk into the land, with half-metre-high walls encircling the perimeter of each to create growing pools. The growing pools are filled with water from the natural spring through a network of narrow canals. Visualise an aerial view of land that has been transformed into a patchwork of fields through the symmetrical use of hedgerows, some fields larger, some rectangular while others are square; the same has been created here through the use of stone, concrete and water.

Outside the packing shed, Ed comes out to greet me, smiling. Ed has close to seventy 'waterboxes': beds filled with the most delicious, refreshing spring water I've ever tasted. The water flows at sixty litres per second through the pressure differential created by the difference in altitude between the South Downs, a range of chalk hills roughly thirty kilometres away at its highest point, and the depth of his boreholes, which range from twenty-four to sixty-four metres, depending on the location of the hole.

As rain pours down, it hits the topsoil and then ventures down into a layer of clay. From there, it continues deeper into a layer of chalk and then flows downhill through sheer gravity, until it reaches the sea. To reach this water, they tap down deep with a corkscrew-like apparatus until they reach the layer of chalk. Water flowing through the chalk rises as the difference in pressure changes; just like the hosepipe trick where as you lower the higher end of a water-filled pipe, water springs out of the higher end. The water is rich in both phosphate and nitrogen, but mainly phosphate, which watercress adores. It feeds the springs that in turn feed the network of rock-enclosed beds. Back in the 1830s, there was a mill operating and much of the infrastructure that Ed farms on today is the same; the spring and its channels that guide the water were all there to power the mill. They realised that the infrastructure was ideal for growing watercress and, in time, knocked down the mill, retained the waterworks and formed the beds that are all still in use to this day. The sound of spring water flowing constantly, always at the same pace, is as much a part of this place as the visual richness of the beds, canals, moss-covered rocks, wild primrose and primula flowers. It is truly enchanting.

The watercress is grown by first seeding the concrete floor of these stone beds and then opening the mouth of the canals to enable water to enter. Blessed by the water that flows from the South Downs at a constant 11°C (to be precise, between 10.5°C and 11.5°C) 365 days a year, the watercress sets roots onto the floor of the beds and feeds on the mineral-rich water. The constant water flow and steady temperature are both vital as they give the plants a consistent, healthy environment.

Generally speaking, the taste of watercress is rooted in its mustard oils – these oils are what give it the spicy heat and crisp brightness that are central to its flavour. Naturally potent, mustard oils aren't actually hard to drum up, but their heat needs a counterweight, which is why I look for notes that round off that core intensity of flavour. What makes Ed's watercress truly phenomenal is its complexity, its vigour. This comes from the mineral-rich spring water and the way in which it is grown. The variety of watercress plays a part, but it's the growing practices that build on this complexity and give the plant its robustness. Slow growth allows the watercress plant to develop into a mature plant with dense texture and robust stalks, vibrant in its pigmentation and buttery, shiny leaves. Watercress is mostly a question of pace: slower growth marks the difference between a plant showing clear signs of vitality, and the thin, pale-stalked plants that are widely available but can barely hold a light dressing without wilting.

Modern watercress farming focuses on reducing the time from planting to harvest – speed. The faster the better as it drives down costs but, as nature will have it, flavour suffers and so does nutrition. As with other hardy plants like rocket, which can be harvested multiple times, the longer you can keep plants going the better they will taste, as each time you harvest a plant it comes back with more vigour and character. Modern industrial farming struggles with this, preferring quick successions of new planting resulting in weaker plants with less taste, less character, less structure and less nutritional density.

Much has changed over the last forty years, in part due to mechanisation as well as market demand, which pushes for the year-round production of watercress. However, 6 June remains a seminal day here, marking the day when the watercress – which has been growing for months – goes to flower as it seeks to reproduce. Maximising harvest up to this day is essential, and then it is time to clear the beds and begin harvesting the seedlings that were planted in April, providing summer watercress from June through to September.

Steve, who started working at Hairspring for Ed's father over forty years ago, has probably harvested more watercress – by hand, that is – than pretty much anyone on this planet. He vividly remembers the break in the season and the hard labour involved in clearing out all the beds with sheer muscle power and grit. Back then, a wheelbarrow and fork were as mechanised as it got. In order to prepare the beds for replanting they need cleaning. Using a pitchfork, they rake up the old plants into the wheelbarrow and push them out where they can be composted. This backbreaking work takes time – months, in fact.

For this simple reason, growing in the summer wasn't possible. Yet it is not only mechanisation that created this seasonality: the flowering process naturally dictated the limits of watercress season, we therefore consumed and selected seed from watercress that likes the cold. Ed confirms that plants that have overwintered always taste best, as do ones that have seen a lot of wind in the days before harvesting (this also builds character), and those that have seen abundant rain, as it crisps them up. Watercress builds up the mustard oils through the winter and therefore the peppery flavour that we associate with this plant is at its highest after the colder months.

To plant watercress, there are two methods Ed uses to seed the beds. The first starts the same way as most growing does, by sprouting a seed in a small block of 'soil', usually peat inside a polytunnel. Once the seed has germinated and a very small plant is growing (Ed's watercress seeds take roughly fourteen days), they are ready to be scattered onto the concrete beds.

The roots take to the bedrock and develop from there, with Ed adjusting the water level in tune with the stage of development. The second and less preferable method, but one that relies on no polytunnel space, is done by spinning a bag of peat in a cement mixer with four grams of seed that has been primed with moisture so it has started to split. Once it is well combined, this mixture is scattered on the beds, and from there seed germination takes place.

The reason bedrock is used to line the growing pools rather than soil is purely a result of mechanisation. Previously, when pitchforks and wheelbarrows were used to clear beds, they were thick with mud, but the introduction of diggers and other machinery to clear the beds meant a solid floor became necessary.

Watching the different stages of maturity in the various beds gives a real sense of the developing stages of watercress and how slowly it grows, as the constant flow of water feeds and nurtures it. What starts as small, dainty watercress plants with two to three actual leaves gives way to denser, more full beds as the plants develop more structure, thicker stalks and offshoots with more and more leaves, eventually covering the bed in a carpet of lush, deep, dense, ashy olive green. The bedrock disappears entirely, leaving only watercress as deep and wide as the pool.

When everything goes to plan, you can get six flushes out of a bed. This means that once the plant is ready to harvest, the first flush, you cut it leaving the root and an inch or so of plant, which will then regrow into the second flush. Ed harvests, feeds with a bit of treated chicken manure to reintegrate phosphate into the crop and waits for the next flush to harvest again. Hairspring is completely organic, using no chemical pest controls whatsoever. As a result, Ed must deal with caddis larvae, an aquatic insect that feeds on the young, white tender roots of the watercress plant. The removal of these small roots hinders the plants and, ultimately, they perish. Yellow spot also causes problems – this virus is a fungal infection that arrived roughly fifteen years ago and can wipe out a crop, as the yellow spots on the leaves render the product unsellable even for processing industries. There is no cure at Hairspring for it, just as there is no cure for mosaic leaf, another fungus that marks the leaves. The longer you harvest for, the more flushes, the greater the risk of disease or pest infestation, like blue beetle for example. Yet there are benefits to taking these risks. Older plants develop in flavour and have a higher level of mustard oils, to the point that it creates purple veining along the stalk and on the underside of the leaves. When I look at watercress, I look for these purple streaks and subtle ridging as well as an unmistakable vibrancy that is synonymous with freshness. Often overlooked, these stalks have a flavour that is even more complex than the leaves themselves.

As I wander down the narrow banks to go from one bed to another, there are bricks along the path here and there, seemingly out of place. When the very cold weather comes, which is -3°C or less, Ed and Steve rush out to protect the plants. The bricks are positioned above gaps in the bed's edge wall where the water feeds into the following bed. By placing a brick within the corresponding opening, effectively blocking it, the water level rises. They then flatten the watercress plants so that they are entirely submerged in the water, which, being at a constant 11°C, protects them from the frost. They can withstand being underwater for two days, not much longer, and once the brick is removed and the water level drops, they magically straighten up and continue their growth.

To me, Hairspring has always been emblematic of both cultural preservation and commitment to quality. It also carries a third value, which is the artist's touch. Harder to quantify, this is what dictates what new levels of flavour can be reached once the right seed is planted in the right soil. The farm always seemed to me like a beacon of hope, a model of what is possible, yet Ed has sadly taken the decision to stop farming and end his family's legacy that dates back over one hundred years. Ed's story should serve as a reminder of how important food choices are in supporting farms that produce food that both nourishes us and does right by the planet. Its watercress feeds and thrives on the water flowing constantly and uninterrupted, and in the same vein Hairspring is a source of inspiration for the food system we envisage.

Back to the Wild

Wild foods are the antithesis of industrial agriculture, and they can teach us a thing or two about water, nutritional density and character. We should find ways to mimic growing that occurs in the wild, to use it as a beacon.

Over the last ten thousand years, since we began farming and domesticating plants, we embarked on a selection process that has removed some of the nutritional benefits of wild plants in lieu of more favourable characteristics for human consumption like sweetness, texture and size to name a few. As we think about character development in plants, we can see that the more we protect and safeguard plants, the easier life we provide, the less character they build. Grit is earned on the street not behind a screen. Wild food has one hell of a fight to endure if it is to survive, and this contributes to its outstanding nutritional density and real depth of flavour.

The fight happens above and below the soil. Coastal food like wild sea kale (you can guess who its next of kin are) and wild sea beet live in rocky, sandy soil that is hard to hold on to. Water is intensely saline, and they get battered by the elements as they hold on for dear life on the edges of the coast. Below ground, they have to search hard to find the vast array of macro and micronutrients, in turn building more character and enabling them to find more nutrition the further they explore. Contrary to an irrigated plant, which receives all it needs by drip irrigation right where its roots are. You can start to see the techniques that Romualdo uses to stimulate wild plant behaviour, or the concept of instigating plant stress which mimics the wild environment.

In the tropics, a plant can have periods of intense sunshine with zero rainfall for four weeks in a row and then get pounded with rain for two consecutive days – how do you think that compares with someone who is fed regularly like clockwork for breakfast, lunch and dinner? I've had the pleasure of experiencing bananas indigenous to an island south of Rio de Janeiro; they are about half to a third the size of a Cavendish banana, the main cultivar consumed worldwide. These bananas grow wild and only take in whatever water nature rains on them, and they are incomparable to the bananas we know. They are yellow inside, a pale golden yellow, with tons of acidity to balance the sweetness, whereas the Cavendish has zero acidity. With beautifully dense texture due to its low water content and genetics, you'll never want to eat a supermarket banana again in your life.

When you taste wild food, you're tasting the pinnacle of flavour. When you compare the 'parental' wild plants like sea kale or sea beet with the modern domesticated varieties, they provide a wonderful illustration for the character I speak so fondly of. You will find that everything about them is more intense, more interesting to the eye, more beautiful to the spirit. You will recognise the weathered look, leathery skin and sparkly eyes, and a handful of scars. You will recognise that those physical manifestations of life are not there by coincidence, they are there by circumstance and by choice. And when it comes to food, when taste matters, that circumstance, that choice, is what counts.

CHAPTER FIVE

Local versus Cultural Preservation

There is an interesting interplay, if you look at the root of the term 'agriculture', between the idea of the cultivation and care of our land and the expression of culture. The word agriculture stems from two Latin roots, *ager*, which means 'field' and *colere* which means 'cultivate', but *colere* has other related meanings: 'to tend', 'to till' (as in tilling a field for cultivation) and 'to care for'. There is deep significance in my view that the word 'culture' derives in fact from the concept of cultivation, which is in turn rooted in caring for the land. Our language contains traces of the idea that producing our food was inextricably bound up in collective behaviours of care, and that culture itself emanates from our first attempts at agriculture from the time when we began to settle.

From Agriculture to Food Culture

There is some disagreement as to when humans began to cook food, but it has been around for at least close to eight hundred thousand years, if not two million years. Cooking increased the efficiency with which humans were able to ingest protein, enabling our brains to grow and for us to evolve from *Homo erectus* into *Homo sapiens*. Our relationship with food dictated our evolution and the role of fire, and therefore cooking, in our collective history cannot be underplayed. The way we prepare and process our food has been a defining feature of our society for the last hundreds of thousands of years, which makes our detachment from the practice of cooking in an era of ready meals and ultra-processed foods (UPFs) even more difficult to understand.

The advent of agriculture might have started with survival, but it was guided by an instinct for flavour. We took wild species that were adapted to the local conditions and soil and domesticated those we found tasty and nourishing. We learned to select seeds from the best performing and tasting plants, using only nature's flavours to guide us.

We harvested and cooked these ingredients, which further refined at what stage we chose to grow and harvest them, based on how or what we wanted to eat. We kept selecting for flavour to maximise our enjoyment of food, as food became central to our social lives and took a prominent role in our most cherished traditions. We cannot remove the culinary component from agriculture as it serves a fundamental function in guiding its evolution. As our flavour-first food system developed and our culture flourished with strong regional ties, a feedback loop was ignited to seek flavour and enhance it.

This transition from hunter gatherers to agrarian settlers also set in motion the development of a stronger culture. As agriculture took hold, we took control of our food supply for the first time. By growing our own vegetables and domesticating animals, we now had sufficient, reliable sources of food and protein. The domestication of animals provided valuable agrarian labour that further increased our food production capabilities. Easier access to fresh food and stores of protein negated the need to search and risk our lives; this meant there was more free time, which led to the second important milestone – specialisation of labour. With more secure food resources, there was now time to specialise. This allowed society to advance at a faster pace and, over the course of roughly ten thousand years, we developed into more complex societies, each with a rich tangle of culture rooted in a specific stretch of land. The arts, music, literature, sport, gastronomy – they are all the product of our ability to divide labour across individuals within a society.

In looking for a more reliable source of food through agriculture, we also created the time to develop food preparation into a craft in its own right, creating and combining flavours on plates in new and distinctive ways, and handing these thoughts down from generation to generation. This, to me, is one of the most endearing expressions of the human species.

The Milpa System

In Mexico, and much of Mesoamerica, there is an ancient tradition that dates back close to seven thousand years of companion planting called milpa, and it 'is one of the most successful human inventions ever created.'[1] The three core crops of the milpa are corn, beans and squash and they are planted together as they benefit from each other. One of the most beautiful connections is how the beans utilise the tall corn stalks as a natural trellis to latch on to and grow upwards reaching for the sun; in turn, they put nitrogen into the soil, which the corn needs. Squash plants have large leaves and spread vigorously across the ground providing shade and retaining moisture, a really valuable resource as small farmers even to this day cannot afford irrigation. Nutritionally, these three crops provide for a super complete diet – writing this leaves me a little speechless and then I wonder how common sense in these matters of our health has been removed from our current food system.

Beyond the agricultural benefits of the milpa as a farming system, it has influenced gastronomy. The consumption of corn, beans and squash is as culturally ingrained as potatoes are to the Irish or olive oil to Mediterranean communities. These are profound evolutionary milestones that we must realise to reconnect our food system to a more harmonious existence with nature.

Remnants of the Past

Food stands shoulder to shoulder with language as one of the strongest emblems of a community's culture. Immigrants will not always pass on their native language to their children, sometimes deliberately, sometimes not, and this allows them to integrate more quickly, but the comfort of home is never far away when it comes to food. This is a very common occurrence in the US where many Asian and Latino immigrants have children who seldom speak the mother tongue of their parents, but who are all too familiar with the food traditions that have been passed down to them. It is easier to change the way you speak than the way you eat.

This is because food memories win out over acquired tastes: food is tradition, and that is worth holding on to. John S. Allen, author of *The Omnivorous Mind*, states that 'Finding food is so important to survival that it is clear that the hippocampus is primed to form memories about and around food.'[2] This essentially means our brains are designed to capture and retain food memories. The hippocampus is the part of our brain that is responsible for long-term memories; interestingly it also holds hormone receptors for appetite and digestive functions. This is why childhood memories are inextricably tied to our idea of flavour throughout our adult lives. The brain recognises the taste and smell of specific dishes, and this has the power to take us back to that moment in an instant.

Proof of this, fortunately, is still right in front of us. We have remnants of a much older, sounder food system residing within the new industrialised one in the form of seeds and growing practices. A handful of families are holding on to varieties that have been handed down to them and continue cultivating them, unseen and unrecognised, whether it is out of pure joy, a sense of nostalgia or just stubborn determination. Fortunately, the industrialisation of our system is decades, not centuries old, and there is still time for us to imprint real flavour into the memories of our children. As my mother pertinently said to me the other day: 'Once you get used to eating good food, it is very hard to stop.' Preserving and growing the market for these remnants, those flavour-first varieties and the traditional farming practices, becomes more and more important as the years pass and our window for passing down this knowledge closes.

Where hope lies is in creating memories. Our deep-rooted culinary traditions can still be salvaged from this mess. Hope is present every time we come together to celebrate. In cultures around the world, most of which revolve around food, we continue to celebrate alongside food, and I find that terribly encouraging. Our joy of eating still runs strong; only our sense of flavour has been distorted.

Where there is a strong connection to nature, a strong connection to food will often follow. It is no surprise that in cultures where a wide variety of plants are still being farmed, I see children eating a more diverse range of foods and fewer processed ones. I seek out regions where farmers grow for diversity and for flavour as this is also a place where strong flavour memories will still be alive. The key is then to bring their produce to places where those memories might be missing, and where that agricultural breadth has been eroded: both need to be rekindled in tandem.

It starts at mealtimes. The most important moments in our lives are usually marked by food, and none more so than the family meal. In those regions where food is less meaningful, it is common for family members to eat at different times. This is heresy in China or Italy. It is unfortunate and not surprising that in countries where this tradition has largely been lost, like in the US, culinary tradition is weak in comparison. The loss of connective tissue that sharing a meal provides is significant. A family needs that time, that ritual. The sharing of knowledge, the passing on of history to younger generations, education, formalities and simple things like table manners and the development of a diverse palate, all take place around a table.

Radicchio: A Story of Survival

In a little town in southern Sicily, you can walk into the local café and find three, maybe four, generations of growers having a mid-morning espresso. The likelihood is they all farm the same varieties of tomato that are emblematic of the locality and feature prominently in the town gastronomy and farming ecosystem. The cumulative experience in that café will be easily over one hundred years and closer to one hundred and fifty years. Add to that what the eldest gained from the sixty years of espressos with his elders who are no longer there – this is what we need to preserve and protect. From a flavour perspective, it is terribly hard to compete against this depth of knowledge and the gastronomic traditions that keep it evolving. The greatest flavour is found where the well of knowledge is the deepest.

Radicchios are a great example of this transfer of generational, product-specific knowledge and the threat posed by recent industrialisation. Grown in the Veneto region in north-eastern Italy, these unique plant varieties did not land there by pure chance – some farmer didn't flip through a seed catalogue and take a punt. There is a history, a soil and a climate ideal for its growing.

There is a closed loop that constantly feeds on itself, between the agricultural craft, the culture of cooking and eating, and the refinement to produce better flavour and enhanced cooking and eating qualities through seed selection. It is a constantly evolving natural process that is wholly ingrained in that culture, and hence why it is of such importance not just in terms of flavour but also in terms of human heritage. It absorbs everyone in the community, from the farmer to the home cook to the wider community that consumes it. The locals have the palate that appreciates its delicious bitterness, and there is a culinary expertise and tradition that has helped steer the radicchio's development over hundreds of years, including highly complex farming techniques for many of the varieties. A challenging flavour, with complex farming techniques and fiercely regional varieties.

Years ago, I uncovered the most precious forcing technique on a sourcing trip for Natoora – on the brink of extinction, its value is incalculable. I will never forget my first meeting with Emerenziana and her husband, Antonello. In my field of work, it is the equivalent of meeting Modigliani or Verdi – these humble humans that we classify as farmers are as much artists as great painters, sculptors or composers. Truly, their masterpieces come from the same esoteric, deeply emotional place, in profound contact with nature like the Masters themselves. Yes, we consume these works of art, but that should not mean we see them as merely food and not recognise the outstanding human beauty at work in the same light we appreciate Monet's brushstrokes and colour balance. The ability to capture nature as Monet did requires deep sensibility, serenity for patient observation of detail, mastery of craft and a stroke of what we term genius. Yet what about the woman who uses nature as her canvas, crafting with plant life and creating something from nature that will then be captured for posterity; is that not art?

I had never seen a pink radicchio-forcing shed before in my life. In the local dialect it is called *radicchio di busa* (radicchio from the pit or cave) because the sheds resemble a dark cave. Inside, there are hundreds, maybe a couple of thousand, of tightly packed plants, so tight they create a thick carpet, completely obscuring the soil below. The soil is actually sand, the ideal ground for forcing them. This technique would now be extinct were it not for the efforts of no more than perhaps two growers in the region, which is to say in the world.

Back in the day pink radicchio, and the more popular Radicchio Castelfranco, was all forced this way. Varieties were selected to undergo this process to produce a much sweeter and crisper radicchio than those grown using more modern techniques. As with all forced plants, they are first grown outdoors and, when ready, they are transferred into a forcing environment, usually a dark shed.

Modern varieties have been developed so they colour on the field, known as self-blanching, to avoid any need for forcing in a dark environment. They are harvested outdoors; their outer green leaves are removed and they are ready to go to market without the need to force them. An additional benefit of avoiding the forcing process, aside from the obvious reduced time and labour, and lack of skill needed, is increased shelf life.

When radicchios come from the field into the forcing shed, they arrive with many rotten and green leaves, and a process of regrowth begins, a regeneration of sorts so that most if not all of the edible part of the 'new' radicchio has grown in darkness and is totally fresh. Because the new growth occurs without photosynthesis, the new leaves are more tender and have more crunch. This crispness, which comes from the water the plant absorbs during this second stage of growth, is fantastic, a true sign of quality, but the added water content does shorten the life of the product, a very noble compromise.

I spoke at length with Antonello about the differences between the modern varieties versus the old ones, which he keeps seed from. He told me that as the popularity of Castelfranco and the pink varieties has grown, seed selection has gone 'the way of the supermarket' – increasing yields, reducing time to harvest and simplifying the process, all to the detriment of flavour and cultural heritage. Flavour has been shoved back in these modern varieties, all to serve the ever-increasing demand, while keeping prices suppressed so they can sit on a supermarket shelf. The tragedy with this obtuse approach is that buyer ignorance prevails, incapable of recognising the difference between a forced pink radicchio and one sold straight off the field. This ignorance is carried through the supply chain and ultimately markets adapt, as they are unwilling to pay a premium for the better product. It is tragic, as it displaces farmers who preserve traditions and know how to grow for exceptional flavour.

In the shed, a young American farmer by the name of Chris was next to me, and all I kept hearing was, 'This is just insane'. For a young grower in love with chicories, especially radicchios, seeing something so rare is a dream. The value of sharing this information is immeasurable and a cornerstone of a healthier food system. I believe in utilising the relationships we've built in Europe with the finest growers to create a new model in the US, where entrepreneurship is rife, yet the culture and tradition is less developed than in Europe. By combining new and old, there is an opportunity to develop an improved model for the future. As we were chatting in the packing shed, where Antonello and his father were diligently trimming and cleaning radicchios over a bin with small, sharp knives, Chris looked at me and said, 'All this must be preserved.' In that instant, I understood from someone else's point of view, arguably a better one, the importance of this little microcosm in Italy. Much of what Chris and his wife, Jesse, are striving for, pushing for every day back home at Campo Rosso, their farm in Pennsylvania, started right here in little packing sheds like this one. Chris and Jesse are two of the best radicchio farmers in the US, and they have made the pilgrimage here to keep furthering their knowledge and gather invaluable expertise in their quest to grow the best radicchio they can. The death of Emerenziana and Antonello's small farm would spiritually be the death of a piece of Campo Rosso, all the way across the Atlantic Ocean.

Preserving these traditions is not as perfectly easy as it may seem. Or, let me put it this way: the medicine isn't. The cure is simple and precise: pay more money for the product. Of this there is no doubt. It is the rehabilitation that is challenging – creating a viable market that is sustainable in the long term for all parties across the supply chain, all the way to the consumer.

As Emerenziana spoke to me about their family's issues, tears were welling in her eyes. I could see them and I could sense the sadness and desolation in her voice, the feeling of inequity, of dire injustice at the simplest privations her family had to endure while others around her benefitted and profited from far less effort.

This goes beyond a sense of human justice for the preservation of growing artistry; this is social justice, the fabric of our society, how we've established ourselves on this earth. One could argue for market economics and the weeding out of businesses or, as she said, finding something more lucrative to do – but at what cost to society? We are not talking about technical innovations or the weeding out of inefficient factories. This is food, what sustains us; it is one of the greatest pleasures we can have as humans and one of the single most defining attributes of humanity that sets us apart from other animals. Not to be overlooked, it is also our closest connection with the natural world, that which sustains our life on this earth and gives us a sense of place with all the sanity that provides.

I can see a world with driverless cars, but can anyone see a world with one common language where that language has been simplified to, say, 10 per cent of the vocabulary we use today? Does anyone want to remove that cultural richness from our existence? So why would we be OK with homogenising our food, removing all the cultural richness that makes us who we are, that explains our love of travel, our desire to live in different countries, to experience other cultures and enjoy food that is so incredibly diverse? Without culture we are no longer humans. This is why there is social injustice in Emerenziana and Antonello's family story, because their work should be far more valued by society than it is.

You cannot strip this depth of culture out of a product like radicchio and expect to achieve the same levels of quality. Therefore, I'm a big believer that we must find clever ways to ensure its survival. I also value and cherish the preservation of culture. The loss of sand-forced pink radicchio represents to me the equivalent of losing a great work from Picasso or flattening the Colosseum.

How do we build a system that sustains growers who are preserving valuable cultural traditions? Our love of storytelling is one way we can reach a wider audience. Reaching deeper into the emotional connections we make with food can help us broaden the market for truly special ingredients. Reaching new markets is an intelligent way to leverage the globalised food system to support the kind of farming that is right for the planet. The imbalances that exist in our world create opportunities to safeguard traditions that would otherwise be lost. Unfortunately, the greatest produce I have come across is farmed in areas that economically don't have the means to support the farmers dedicated to using flavour-focused, labour-intensive techniques. Competition from industrial agriculture is present even at the higher end of the fruit and vegetable spectrum, with seed companies developing similar varieties that compete in the market and on grocery shelves but lack in flavour and texture.

The sad reality is that our palates and knowledge are evaporating. Hence the farmer growing radicchios the old-fashioned way has to compete in a market flooded with poor imitations that look the part and cost less. If we can find new markets, where consumers have the economic means to value the product correctly and allow it to stand out from the competition, I believe it is worth the extra miles to get it there.

The Tamoa Conundrum: Locality versus Quality

Local food is relative: to one person it can be the country they're in while to another it can be food coming from their state, and to yet another it means food within a thirty-mile radius. In each case, consumers will have a very different frame of reference.

The overarching concept of 'local' stems, as most of these labels do, out of good intentions. Asking people to buy local should drive them to a more seasonal offering, which has travelled fewer miles and is therefore fresher and, if pushed further, one could argue has a smaller carbon footprint. Local is also an economic argument, as through simple consumer choice to support local farmers we can stimulate the rural economy and help sustain those businesses that are close to home. Finally, there is the cultural impact on a specific community: if as a result local businesses involved in growing or producing food persevere, so do the regional gastronomic traditions and we can infer that the local culture will be richer for it. But what if the local rural economy is not ready and willing to preserve that valuable tradition? Should we let it extinguish and relinquish it to history?

My experience in building fresh produce supply chains across the world has taught me to always prioritise flavour and product quality. This clarity and agnostic view to country of origin has given us the format to create real impact.

There are numerous cases we can look at where Natoora has stimulated sufficient demand, outside of a product's natural growing region, leading to meaningful preservation of traditional varieties and valuable farming techniques: a tangible form of cultural preservation. With the added benefit that we can reintroduce flavour back into the system in large ecosystems where change needs to happen. In today's globalised world, local food systems are part of the solution but, on their own, they are not powerful enough to stimulate the necessary change to transform the food system. We no longer behave as a local species, and our consumption and eating habits reflect that.

Modern supply chains are fast and efficient and farming practices have evolved – for good or bad. Together, they have enabled a wider (not more diverse, mind you) diet that has shifted consumer expectations. Does anyone really believe we will eliminate limes from gin and tonics around the world if they are not locally available? The solutions must take these realities into account to enact real change. One side of that reality is scale – we cannot shift the system with a snap of our fingers and suddenly rely on small farms to produce the vast amounts of food the world needs. We need diversity in food production and that includes farms of different scale. In today's globalised world, exclusively buying local to revolutionise the system is an impossible dream.

So how should we look at localism? Should we draw a line and, if so, where? These are worthwhile questions we can ask at a personal level. They are the same questions I have asked myself as we've built these vast supply chains in different parts of the world. The work of Tamoa in Mexico helps put these arguments into perspective.

Francisco and his wife, Sofia, started Tamoa as a cultural preservation project – to safeguard Mexico's food heritage by sourcing heirloom varieties of corn, beans and chillies among other native ingredients for restaurants around the world. They have created a market that enables them to source immensely important, and delicious, plant varieties that are at risk of extinction. The parallels with Natoora are evident. Source amazing products with a focus on flavour, work with the best farmers and deliver a product that creates positive impact, supported by education.

Francisco's love of corn, masa and tacos led to him where he is today. I believe he is privileged to be engaged in truly important work, saving the rich history of his country from the industrialisation that is drowning it. In building ethical, flavour-driven supply chains to preserve heirloom varieties and traditional farming practices, he is acting on multiple levels of the local food system.

Francisco and I met through Sebastian Vargas, a Colombian friend and great chef who is a big fan of Tamoa and Natoora, particularly the work we do in highlighting and therefore preserving our rich gastronomic culture. Sebastian was adamant I should meet Francisco or, as he is affectionally called, Uco. After several conversations over the course of the following months, Fede and I arrived in Mexico City on a hot early summer evening.

What awaited us was going to surpass our wildest expectations. Our immersion into a culture so magical and deep proved an impossible gift to repay. And fittingly, it all started with an *al pastor* taco, the classic Mexico City pork taco cooked on a spit like a kebab, garnished with a piece of pineapple, that is burned into my memory. With barely enough time to drop our bags, the first face-to-face conversation with Francisco was over a taco late in Mexico City's night.

I discovered and instantly fell in love with tacos when I was ten years old. Yes, those Old El Paso hard-shelled ones filled with minced beef, shredded lettuce, diced tomato and shredded orange cheese, which I discovered at my American schoolfriends' birthday parties. This crunchy vehicle that would hold all these complementary ingredients together was a revelation, a new form of food. Little did I know that thousands of years earlier, Aztec and Mayan civilisations had created this food, albeit a far less industrialised version than the one I fell in love with. Those crunchy tortillas were part of a 3,500-year-old culinary history. Francisco took my understanding of tacos, maize farming, masa (the nixtamalized corn dough with which you make tortillas and many other foods) and Mexican culture to a level I didn't know existed, and in the process changed my own cooking forever.

I only really understood the magic of this cuisine when, thanks to Francisco, I was invited into the homes of heirloom corn farmers, the real guardians of Mexican culture and tradition. The food those farmers shared with us in their outdoor kitchens was some of the finest in the world. I struggle to find the words to describe the intensity of love and terroir present in the food. It nourished you with mad deliciousness but also vast amounts of intelligence. Everything had place and meaning, every ingredient, like a musical instrument in a well-rehearsed orchestra, had reason to be there. And nothing was there that did not add to the experience. Maize behaved like a mythical god holding court as dried corn ears were burned for fuel acting like charcoal, the kernels nixtamalized and were used to make a delicious soup called *pozole* and a crazy delicious drink called *atole*. (Nixtamalization is a process developed in Mesoamerica to prepare grain for consumption by cooking it in an alkaline solution, usually water with lime, that carries significant benefits as the process renders the maize more nutritious, reduces toxins, makes it easier to grind and improves its flavour.) The nixtamal was ground into masa, which was then transformed into tortillas, *tetelas*, *sopes* (or *memelas*) and crispy *tlayudas*, while dried corn husks held *tamales* tightly in place. Perhaps the greatest thing I ate was a tortilla straight out of the *comal*, the traditional flat, round cookware, which is a staple of any Mexican kitchen. The tortilla is seasoned with rendered pig fat and salt, then rolled into a cigar shape and pressed to create a sort of churro that goes by the name *gordita* – what kids used to eat when they returned from school. *Gordita*: three ingredients, masterfully crafted into simple perfection.

The work these families are performing for society is so invaluable it really is hard to fully grasp. A dedication and commitment to the preservation of seed varieties and farming traditions that embody a whole civilisation. Corn, masa, tortillas, food, family, soil, beans, fertility, nutrition – they all coexist; in a way they are all one. It is the Mexican way of life. However, the quest to preserve native corn in Mexico depends on foreign markets.

To fulfil its mission, Tamoa has to convince other cultures to embrace Mexican culture. How bizarre. Mexico has a phenomenal culinary culture dating back thousands of years, yet it must rely on wealthy diners in major international cities to preserve its heritage. Tamoa barely sells any corn in Mexico City, as very few of the city's restaurants want to pay the price for upholding tradition and putting flavour first on their plates. It is a sad state but, were it not for the flexibility to go beyond localism, there would not be enough support in its own territory for Tamoa's mission. So, Tamoa taps into the international markets where chefs in major international cities like New York, LA, London and Copenhagen have the will and clientele to uphold such a worthy cause. When I hear individuals who say they don't believe in shipping product across continents, who in the globalised world of today continue to press for exclusively small-scale, local food systems, I always think of the Mexican quagmire.

The industrialisation of agriculture has reached Mexico. Even they have not been able to hold it back and it is having real, negative consequences on their heritage. Mexico, the birthplace of corn, today imports vast amounts of yellow corn from the US, which is genetically modified, and it has the second-highest obesity rate in the world, just behind the US. Mexican life is so intertwined with maize and masa that it is unthinkable to see old traditions extinguish, yet this is precisely what is happening, with the exception of the efforts of Tamoa, as well as Masienda, another great organisation doing the same vital work. And thanks to the international community who believe in the importance of this work, not just for Mexico, not just for flavour, but for contributing to the preservation of the precious human heritage we all share.

The Local Debate

The often-touted advice that you should eat locally to save the planet is extremely misguided. Most of the greenhouse gas emissions associated with our food are generated by its production, not through the transportation of such food. It has been shown in recent years that 'the carbon footprint of transporting food is relatively small, and that it's more important to focus on how your food is produced. Eating local can be a part of that, but it doesn't have to be.'[3] In other words, what we eat and how it is grown is far more important if we're making dietary choices based on CO_2 emissions.

I believe localism should be looked at through a seasonal and cultural lens. Distance and in turn freshness is a great barometer of where to draw the imaginary local boundary. Transport systems, existing supply chains, road infrastructure and even the size of the country or region should all be taken into consideration. Getting product from southern Sicily into Paris is easier and faster than moving product from Scotland into London. Moving dried corn and chillies from Mexico to New York is relatively effortless and is more than justified by the valuable work it creates for farmers all around Mexico.

At Natoora we don't see seasonality within the constraints of UK weather patterns, which limit what can be farmed in different climates further south. If we can source well-farmed and delicious vegetables from southern Spain, we believe in supporting those growers and broadening the market, particularly when it takes little time to move products within Europe. If you guide your sourcing with flavour at its core, it can be a powerful force for good. Flavour really precedes seasonality, in the same way that I've always argued for the concept of 'I eat, therefore I'm seasonal'. Superb flavour and quality can only result from products grown in their natural season.

I place a great deal of significance on local heritage for I believe that it elevates the levels of flavour and quality possible. The concept of craft as 'an activity involving a special skill at making things with your hands'[4] evolves through time via a hands-on transfer of knowledge and expertise. Crafts by their nature cannot be acquired exclusively with an academic transfer of knowledge. You can study how to grow tomatoes, but you will only learn how to grow fantastic tomatoes by learning from other farmers with more experience along with years of dedication to the craft. Local heritage is simply that – an accumulation of knowledge specific to a craft that is ingrained in the local community, like the growing of radicchios or the milpa farming system.

Tomatoes for me represent the best dichotomy in the local versus quality dilemma, and one that tackles the food miles and environmental impact argument head on. The misconception stems from apparently rational thinking that buying local equals a lower carbon footprint. In the UK, there is a brand of tomatoes that markets itself under the local label. Many consumers, understandably, buy them as a way to support UK farms, which is something I entirely respect. Those 'local' tomatoes grown in the UK actually have a higher carbon footprint than those grown in Sicily because their farming requires artificial heat. Greenhouses in the UK need to be heated regularly, which is not the case in Sicily. In Sicily rarely, and only on the coldest winter days, do they need to heat greenhouses overnight to avoid loss of crops. In the UK, even during the summer months, greenhouses are regularly heated to ensure consistent growth to foster commercial production and keep tomatoes on supermarket shelves. A study in Austria sheds some interesting light:

> Our results show that imported tomatoes from Spain and Italy have two times lower greenhouse gas emissions than those produced in Austria in capital-intensive heated systems. On the contrary, tomatoes from Spain and Italy were found to have 3.7 to 4.7 times

> higher greenhouse gas emissions in comparison to less-intensive organic production systems in Austria. Therefore, greenhouse gas emissions from tomato production highly depend on the production system such as the prevalence or absence of heating.[5]

As we can see, heating is a major contributor to greenhouse gas emissions thus increasing the carbon footprint, and all for a product with significantly poorer flavour. The impact of trucking from Sicily to London is lower than people think. You need more manufactured inputs in regions that do not have the right climate.

Additionally, tomato plants need sunlight, and lots of it, along with the sun's heat. Flavour is dependent on that sunlight, as is the soil, which needs to be minerally rich. These UK tomatoes are hydroponically grown, not grown in soil. The only reason they are English or local is that they are farmed by a UK-registered business whose greenhouses happen to be on UK land, even though the roots of those plants never touch UK soil. Those roots most likely grow on crushed coconut husk from the Far East and are fed imported nutrients by a computer system that is taking daily readings. Dutch hydroponic growing technology based in the UK with no local heritage being drawn into those plants. You can build that greenhouse anywhere in the world and get very similar results.

I don't believe in this kind of local food system, and the lack of taste is enough for me to source elsewhere. I would much rather stand for culture and tradition, for soil health and enriching the planet's resources, rather than stubborn localism. We cannot separate our food production ecosystem from the culinary traditions and the culture they are a product of. When we think about fixing a system that is so broken, we have to do it by examining tradition, respecting it and using it as a starting point for progress. When you break the connective tissue that holds different components of the food system together, you begin to lose richness, joy and flavour. You also lose some of our humanity. The taste of an incredible plum or an heirloom corn tortilla are far better indicators of the values our food system should aspire to.

CHAPTER SIX

Creating a Market for Flavour

As I think back to my time in New York in the late 90s when I actively began to engage with the lack of quality produce available in large cities, my idea was always to figure out a way to bring flavourful produce to the consumer. Instinctively, and wrongly, I believed restaurants had access to great produce and it was purely a matter of making it accessible to the regular Joe. The consumer is hard done by via the system; even back then, restaurants always had better sources for quality ingredients. Without any experience to draw on, I thought the solution was relatively simple. I had no idea of the realities of the system and the significant challenges it imposes on the problem I wanted to solve.

One Christmas many years ago, I received a call that has stayed with me since for its kindness and wisdom. The man who called was Stefano Fatarella, friend and Tuscan chestnut supplier based in the elegant Monte Amiata. On that day, Stefano made me realise the importance your customers have in creating a market for flavour. Demand, it seems, is the most critical piece of a supply chain: it defines it. And while you can stimulate demand, coax it, market at it, it has the ultimate say on the quality of product and price point that the supply chain can move. What I had not realised until that phone call was that it didn't matter what you wanted to sell to your customers, what mattered was what your customers were willing to buy. His customers, he told me, were not interested in higher-quality fruits and vegetables so, no matter how much he wanted to buy the same products we did, he was not allowed to. With sincerity, Stefano spoke of Natoora's sourcing and urged me to reflect on its integrity. In doing so, he sparked a change in my thinking, opening my eyes to the importance of the customer in addressing change.

Market Realities

Food is perhaps the most complex supply chain system in the world and one of, if not the, largest systems in the global economy. This system moves highly perishable, extremely delicate goods, many with only a handful of days of shelf life, requiring fully refrigerated transport, storage and distribution, and it must be cost-efficient given the relatively low-value nature of the goods moving through it. The catch is that a supply chain of this nature is like plumbing: it is hardwired into our global infrastructure. Of great significance, and to come full circle, supermarkets own the vast majority of the last mile. Those physical locations, where most of our food is accessed, are as hardwired as they come. Imagine deciding that you don't like where water comes out of your home and you just want a tap in the middle of the living room floating in midair; well, for the moment you must run a pipe as we have no other efficient way of delivering water. Food might not appear so constrained, but that is exactly the reality we are in. In developed countries like the US and the UK, more than 80 per cent of food is sold through supermarkets. An entire global food system has been engineered to move product efficiently from farms and food manufacturers through roads, warehouses and factories into stores so that you can walk out of your house and buy a fruit salad at the supermarket. These existing constraints impose challenges that define how you can tackle the food system's issues.

Scale is one of those constraints, and this is more evident in a vast market like the US than in Europe. The industrialisation of agriculture took on an enormous size in the US, larger than anywhere else in the world, and with it came a transportation and distribution system geared towards moving large amounts of food. This means moving small quantities of food becomes increasingly difficult and costly. Of all constraints, scale is the hardest to overcome for a young company like Natoora. When you need only ten boxes of oranges, it is extremely hard to move them from the farm to your warehouse and it creates a chicken and egg situation – you need product to grow your business, but you don't have the volume to justify getting your hands on the product because you need more volume. In the early days, my energy was focused on fixing the problem of access to quality food for the consumer, not for restaurants. But what happened, a little by chance, quickly became a great way to create volume in our own immature supply chain. Restaurants were the perfect partner to build volume, and chefs proved to be instrumental in more ways than one, devoting time, excitement and purchasing power to the cause. This is how I internalised the massively important role that chefs had to play in fixing our food system.

The Role of Chefs

The time I spent in New York after university coincided with the new era of TV chefs and the rise of the chef as celebrity. Today, we all know how influential chefs have become in leading and shaping food trends. That influence has spread from our TV screens, and a handful of chefs, to a far wider catchment of the chef population and also moved from the TV to their actual restaurants and social media platforms.

With the number of meals eaten out of the home rising, an increased interest in food and the celebrity chef's rock star status, it paved the way for the industry to have a lot of sway in shaping consumer choices. My experience is that many chefs, particularly those without a traditional platform like a food show or successful book, don't appreciate the attention they command. Younger generations are eating out more, consume less TV and sadly read fewer books. And if they're eating out more, you know they're using their kitchens less.

Ingredient-led restaurants are having a disproportionate influence on our food choices. I have seen this over my years in the industry. The disproportion is fantastic and needs to be leveraged even more. Chefs, collectively, have a far greater ability to enact change than they individually believe. The MAD academy in Copenhagen has recognised this and works on 'internal to the industry' change through education.

Kitchens exert influence internally as much as they do externally on the wider public. All those young cooks will be the influencers of tomorrow, and the earlier in their careers they hone their buying skills and see what a restaurant is capable of within a food system revolution, the better. The purchasing power I touched on earlier is in part to do with volume. But it also, importantly, touches on price. Ingredient-led restaurants have a special clientele that allows their kitchens to spend on quality produce, to seek out flavour. As a consumer, if I experience great flavour, irrespective of where, I will be more inclined to want to repeat that experience. Flavour-driven positive reinforcement is how we change what we ask of our food system.

Restaurants are ideal locations for consumers to taste diversity and experience fantastic flavour.

From a supply chain perspective, if we look at the inner workings of the system, ingredient-led restaurants have always been seeking out better and better flavour. Those kitchens understand the importance of great ingredients to their own execution, and the simpler the menu, the lighter on technique the cuisine, the more this is true. I have always wanted to find the best artichoke in the world, but without the support of the chef community I honestly would not have been able to. There is an enormous gap between identifying an incredible peach, for example, and identifying an incredible peach and making that available to an enormously wide audience.

A supply chain by its very nature needs to move product from point A to point B. The chef's natural alignment with flavour above all else is what allowed me to honour my own vision. And when chefs get behind an ingredient, they have mighty strength. A wonderful example is the Delica pumpkin grown by Oscar Zerbinati in Mantua, Italy, a region renowned for the cultivation of *Cucurbita*, specifically melons and pumpkins.

Known as the Delica in Italy but originally from Japan, this pumpkin is more widely known as Kabocha and grown around the world. But there is a key difference in Oscar's Delica – a post-harvest curing process that removes moisture and radically improves flavour. Placed inside a heated shed, the pumpkins lose up to 30 per cent of their weight, mostly water, concentrating sugars and flavour. Oscar is mimicking his grandfather's technique for the pumpkins he would save for home. The results are out of this world.

Standard Kabochas average six to eight Brix, which is a measure of the sugar content. For comparison, a very good Datterino tomato is around twelve Brix, while a Reine Claude

Dorée can reach more than thirty. Oscar's pumpkins reach fifteen! The result of this commitment to flavour, and the strength of the chef community, can be seen on menus across London both in restaurants and home kitchens alike. A decade on from first introducing it to the London market, the Zerbinati Delica is likely the most widely used ingredient across kitchens during its autumn and winter seasons, and thousands of people in the UK now enjoy this incredible pumpkin at home. A chef's collective power enables a supply chain for that specific flavour to emerge through the menus they create and goes on to influence thousands of consumers, journalists and foodies.

Responsible chefs and the kitchens they lead are shaping our culture, by acting on food and guiding its evolution. Restaurants committed to flavour, that source with integrity as best they can, can leverage their outsized influence for real impact. It is a very powerful flywheel. The story of winter tomatoes is the example I'm most proud of.

The Camone

My mother introduced me to the Camone tomato in the 1980s. We spent a handful of years in that decade living in Italy. The salty, crunchy skins of the Camone, with their perfectly round body and dark green shoulders, were instantly appealing to my tastebuds.

The Camone is a Sardinian tomato, and I like to think local growers created it through the terseness of Sardinian character. My sourcing travels brought me to this island to find the real story of this amazing fruit. Islands always have a unique feel to them – there's a climate and a culture that belongs to them and only them. Islanders are insulated, protected in some ways from external influences and, as such, they and their customs attain a level of integrity, which is extremely special. Sardinia, just like Corsica, would prefer to be a nation state of its own; nonetheless, the Italian influence is present, although maybe the influence is Mediterranean instead.

The story of the Camone will never be definitive; as with much of our recorded history, it is highly dependent on the point of view of those who experienced it. This is the story I pieced together from conversations with local farmers. The area is Pula, on the southern coast of Sardinia, about thirty kilometres west of Cagliari.

In and around the 1980s, some very talented and innovative growers decided to play around with a tomato with the name Camone by changing how it was grown, which resulted in a new and better Camone tomato. It is a readily available seed that produces a rather uninteresting and quite common round salad tomato, mostly red with a slightly greenish top and of normal size. A Dutch agronomist from one of the large seed companies apparently developed the variety. What is certain, as many parts of this story are not, is that a Dutch agronomist was shocked when he came over to Sardinia and saw how farmers were growing the variety he developed. Even in the friendliest of circumstances, imagine how difficult a meeting of minds between a Dutchman and a Sardinian can be. Both have arms and legs and walk upright, but their minds could not be any further apart. Out of this juxtaposition comes a story of great beauty.

These rebels, these unwieldly Sardinians, had broken all the rules. They had gone against all the recommended practices the seed company had in place for this variety. Importantly, they planted the tomatoes at the wrong time of year. Seeds were planted in autumn, in sandy soil a few hundred metres from the sea, and the plants were stressed, significantly so. Partly, this happened a little by accident; Pula, being close to the sea, has naturally high-salinity water and the lax attitude of the growers who, caring little for plant vigour and productivity, magically came together to produce fruit that was far smaller, incredibly flavourful and with a deep green top.

This new fruit was a stark contrast to the tomato as we know it. Crunchy instead of soft, thick skinned instead of thin, salty instead of sweet – but boy is it addictive. This was the birthplace of this tomato typology, or so I believed. My search for the truth continued as my obsession-fuelled interest in these tomatoes grew.

The Marinda

In January 2010, I was on one of my regular visits to the wholesale produce market in Milan. It was blistery cold in the early hours before the sun rises. But the energy is all consuming and keeps you humming. Seeking out produce is infectious even in the confines of a market.

The first time I set eyes on a case of Marinda tomatoes is a sight I won't forget. The attraction was instinctive. Its physical traits so alluring, its ridges and colouration so rich in character. I inspected them for a while before taking a bite – they reminded me of a ridged tomato called Riccio Fiorentino, similar in size and shape but fully red and soft. I've never needed convincing of the Marinda's flavour and texture, though it jars with our concept of a fully ripe tomato, and I instantly recognised the physical and flavour similarities with the Camone. I bought our first Sicilian Marinda tomatoes and off they went to London.

In that first season sourcing the Marinda, we realised very quickly the vast range of quality that the variety can produce. To put it into context, if a normal product has a range of flavour that you could grade from one to five, the Marinda's range is easily one to ten. The really incredible-tasting tomatoes represent only a small percentage of all the tomatoes the Marinda plant produces. Their mad flavour and appearance, their similarity to the Camone, this vast range of quality all made me want to know more – I had to dig deeper. Why were these tomatoes so insane, and why were they available in the winter and early spring? I needed answers. In the meantime, I enjoyed the hell out of them, could not stop eating them and built a great understanding of how to treat them.

This tomato would take me to a little village with the most enchanting of names, Portopalo di Capo Passero. Portopalo, as it is affectionately known, is in the southern tip of Sicily and right by the sea. Over many trips, I would learn that the very similar technique utilised in growing Camone tomatoes over in Sardinia was actually taking place in this part of the world in the 1970s. The concepts were identical – really stress the plants using high-salinity water, and little of it, and grow them during the colder winter months.

As much as I loved this tomato, I was finding it hard to buy as much as I wanted because I could not convince enough customers to order them. My absolute obsession with their amazing flavour, and a commitment to seasonality, is why the story unfolds.

The Birth of Winter Tomatoes

It all started in the winter of 2008 with the first Camone, a tomato that was by then already coming into the UK in very limited quantities. We began sourcing it through a wholesaler in Milan with deep connections in Sardinia.

Back then, Santino Busciglio, a Sicilian chef who is super into flavour and simplicity, opened a restaurant called Mennula on Charlotte Street in London. He was one of the first consistent customers of Camone tomatoes and deserves credit for being an early believer. Within a year, the likes of Theo Randall and the restaurant Sketch were buying some, but it was still hard work. And then, by January of 2010, we introduced the Marinda, which I took out on every single customer visit, but to my dismay I could not for the life of me understand why very few customers would order them. Here you had a product with phenomenal flavour and real seasonality!

Back in those days, I used to visit Claude Bosi at Hibiscus quite regularly (another story is how I discovered he hates rosemary) and kept bringing him these tomatoes. He loved them, and he told me so. He also told me he could not put them on the menu because tomatoes were a summer product. He could not have tomatoes on his menu in winter. We had several of these conversations and I would tell him about their seasonality, how these tomatoes were not available in the summer. That you could only grow them during the colder months, the time when you could really stress the tomato plants. This was an innovative farming technique that was producing a fiercely seasonal product with out-of-this-world flavour – it was built for chefs like him.

Those conversations proved invaluable as they got me thinking about the tomato's natural summer seasonality and how to counter it. Claude was not the only chef who had this issue; I was having similar difficulties with The River Cafe, the most religiously seasonal restaurant. I had to find a way to give them agency to put a tomato on the menu in winter. Their reputations depended on it.

Then I had an epiphany! I decided to create a new category of tomatoes by grouping together the Marinda and Camone varieties, two products with distinct regionality but incredibly similar flavour and physical characteristics, with nearly identical farming practices. If I could get chefs to believe me when I told them that these tomatoes were in season in the winter, then perhaps they could convince their customers to do the same.

So that's how one day, in the winter of 2011, I wrote the words 'winter tomato' for the first time in our weekly newsletter to chefs to describe the category of tomatoes represented by the Marinda and the Camone.

As if by magic – and, in reality, through years of perseverance – The River Cafe in March of 2012, at the peak of Marinda season, put them on their menu. Over the years, the winter category has grown to include the iconic RAF tomato as well as the more modern Black Iberiko, both grown in Almeria in southern Spain, and all sharing the same farming techniques, flavour and physical expression.

Two words that until then were never seen together were more powerful than flavour. But this story would never have continued were it not underpinned by what many have described to me as life-changing flavour. As time went on, more and more kitchens began buying them and using the term 'winter tomatoes' on menus across London. From there, journalists got hold of the term, and then our competitors and even consumers. Whenever I see an article referencing them, I feel happy inside and a little proud of how, with the help of the chef community, I was able to coin a term and create a new category of tomatoes. There is no better example of the ability of the restaurant industry to shape what we eat – and that is one step closer to shaping the future of our culture. Chefs have a real role in bringing flavour to the forefront of our food system and in unlocking all the benefits seeking out flavour has in and around it.

Creating a Market for Green Citrus

The story of green citrus is a story about fresh thinking and, while it will not solve the greater problem facing us, it is a beautiful example of thinking differently, as Steve Jobs would say. 'Green citrus' is the name I gave to a new category of fruit:

a selection of different citrus varieties that we harvest in their green state prior to them being fully ripe.

All citrus fruits start life green on the outside. Their skins at this unripe stage have not yet turned yellow or orange but, interestingly, their essence and perfume are more pronounced as the volatile oils in the skin are at their peak and they mellow as the fruit ripens. The same occurs with acidity, which is higher when unripe – as fruits ripen, the sugars become more prominent and acidity likewise mellows.

In the summer of 2013, I was on holiday to the Amalfi Coast – an area of exceptional beauty – during the first weeks of September. It is a golden time of year to travel there; with tourism fading, there is freedom in the salty air. Over lunch at La Tagliata in Montepertuso, I was wandering the restaurant's vegetable patch that overlooks the magnificent coast, where summer fruits abound with aubergines, courgettes and Sorrento tomatoes all growing in open fields at the mercy of the weather.

A pergola, as is customary in the area, was providing much-needed shade over evenly spaced lemon and orange trees. As I walked around the trees, I looked at these oranges, still very green, hard to the touch but so enticing, and something clicked as I stood there. My mind took me to bergamot, a citrus fruit I had been cooking and playing with for some time. I understood their culinary applications at this unripe stage were greater and more interesting than with their ripe counterparts. Their magnified fragrance and perfume in their green state, their juice full of complexity, and aromas and volatile essential oils at the height of their activity – these are remarkable qualities that took my mind into a world of possibility, and to a very special langoustine dish.

I had never seen a fresh bergamot before I came face to face with one at the wholesale market in Milan. Bergamots were known to me – their essential oils are used extensively in the perfume industry, where it is the primary aroma in men's cologne, and it is the distinctive flavour of Earl Grey tea. Then, it was rare to find fresh bergamots as the entire production, of which Calabria in southern Italy commands 90 per cent globally, was earmarked for making essential bergamot oil.

When those first boxes arrived in our London warehouse, I paid closer attention, cut one open and was floored. Flavour and fragrance unlike anything I'd experienced before. In that moment, I could think of no better chef to go and see than my dear friend Antonin Bonnet, who at the time was heading up the kitchen at The Greenhouse in Mayfair. I adore his cooking and have the fondest memories of spending countless hours in his kitchen talking, tasting, eating, usually on my way home at the end of a hard day. His sensibility for produce is one of a kind and we bonded over a love of great ingredients. As I expected, he was as blown away as I was, and the next day he served me the finest langoustine dish ever – the elegant shellfish elevated by the complex floral aromas of bergamot; it was otherworldly. This was the moment, the seed, of what would develop into a new category of citrus fruits.

Back in the orchard on the Amalfi Coast, I thought of Federico as I clipped two oranges, pocketed them and thought of trying one after lunch and the second one back at Natoora in a week's time with him. I had a sense of possibility, of discovery, thinking this might actually work: commercialising an unripe fruit for its cooking prowess, its seasoning capabilities, its complex notes and that infinitely valuable astringent acidity – just think of lemon's versatility with food. This was the beginning, the moment of fleeting inspiration and innovation. This unique way of thinking I believe can engage the chef community, inspire them to create and, if we all work hard enough, allows us to readdress the importance of food full of flavour in our culture.

In the years since Federico and I thought of creating the 'green citrus' category, it has taken on a life of its own. We've been able to replicate it in the US and France, and now menus around the world feature different varieties of green citrus. Early on, we knew we needed to create a category giving chefs variety of flavour and culinary applications. One green variety was not sufficient; chefs needed diversity, which I knew from my own cooking. In season they needed options: one variety whose zest goes onto raw scallop, while another can be segmented and added to a salad. The idea needed relevance and depth to succeed. By creating a category, as we had done with winter tomatoes, we gave meaning and resonance to the product. We gave chefs a vision they could understand and therefore work with.

Fede and I explored different varietals and tasted them, and then picked a handful that were the best but different enough from each other that they could be used in different ways. Fruits that best reflected this new category and that also lent themselves to far more culinary uses than their ripe counterparts because of their heightened aroma and acidity. This category stands as a beautiful, natural reflection of the seasons, their gentle progression and variability. Years later in New York, we went through the same exercise, working with our farmers in California to select a few varieties to build the category. We have a green Meyer lemon, the green Minneola tangelo, a green pomelo and, like in Europe, the green Cara Cara orange.

Cara Caras are a cross between a Navel orange and a pink grapefruit; we've aptly renamed them Pink Navels. In their green state, they represent a stunning contrast of outer green skin and orange-imbued pink flesh, a marvellous contrast of colours. These are picked about one month ahead of maturity, as they need more time to fully develop their juice. The Navel, also picked around one month ahead, is better suited to being picked green than an orange like the Tarocco, which is naturally more acidic. The greater levels of sweetness in the Navel give the green version a more palatable orange flavour, less reminiscent of lemon.

Green mandarins represent the finest of the green citrus that we source, picked roughly two months prior to being fully ripe, and about half or maybe 60 per cent of its final size. Interestingly, its seeds are fully developed and are outsized versus the rest of the fruit, yet they are still very green and buttery. The skin is pine green, vibrant and luminous, its shade of green is super intense and dark; profound. The characteristic small leaves are massively fragrant, while the inside reveals early on a yellowish flesh and later on in this sub-season a bright early orange tone, visually revealing the unbelievable amount of juice already present in the fruit; a juice that is at once acidic, with green undertones and an indescribable aroma of mandarin perfection, a complicated mix of citrus scents carefully brought together. The volatile essential oils are wild.

Carmelo, whom I introduced on page 33 in Seasonality 365, was instrumental in the discovery of mandarins in their green state and in the development of the green citrus category. When Fede and I planned the sourcing trips to develop this category, we spoke to several farmers about our crazy idea. Carmelo was instantly drawn to the idea and willing to make this happen. With a childish flair, as we wandered his orchards, he said he had a variety that was going to blow our minds in its green state. While he had not commercialised green citrus before, ever the perfectionist he had tried fruits at this early stage plenty of times and knew precisely what we were trying to create. This was the first time we tried mandarins off the tree at this stage. Fede grabbed one and gently squeezed the whole mandarin between two fingers; at every squeeze, a mist of oils vaporised from all around the skin surface, like a porcupine fish extending all its spikes and then retracting them, over and over, fully loaded with wild, crazy, mind-bending oils. Shellfish will love it – even a little finish of grated zest will be enough. Crystal sparkling water will be elevated with half a mandarin squeezed in, while greatly crafted gin with its ebullient botanicals will start flying when it comes into contact with green mandarin juice, some ice and soda.

The introduction into our kitchens, our lives, of unripe yet in some ways fully developed fruit further refines the seasons; it slices them into more subtle layers that give new complexity and meaning to our connection with nature and, in turn, the changing of the seasons. If what the supermarkets have done through their year-round sourcing policies is to flatten and remove complexity, creating a superhighway, we're adding bumps and troughs, puddles and forks in the road, adding thought and interest, naturally slowing the road down to a speed appropriate for the real appreciation of nature. If you ever go to the Amalfi Coast, you must experience it by boat. It is the only way to completely and wholly grasp its beauty and essence; travelling in a *gozzo*, the traditional wooden boat used in the Mediterranean, slows your journey to a pace that all of a sudden opens up and stretches the coast like an accordion, exposing all its detail, the crevices, the magnificent nature and exuberance of it. Pace is so relevant and critical; green citrus gives the food system pace.

The real challenge is managing a fast-growing business, aiming for a swift change of the system, while living by the realities of nature and of living plants. The food system of tomorrow cannot scale instantly by adding computing power; this is an industry constrained by pre-existing critical infrastructure. Natural food depends on and is slowed by the natural life cycle of plants. My view is this actually strengthens the system, it grounds it in a physicality that connects us to our planet, one that technology at times disconnects us from. Chefs, however, can move extremely quickly and utilise their craft to be the tip of the spear of innovation and change.

These stories illustrate the role chefs have played in stimulating a flavour-first food system and serves as a playbook for how the global chef community can be a force for positive change. Together, they represent serious purchasing power, and my intention is to have them work in concert to speed up change. Their collective influence on consumers can further amplify the message to demand better, to seek flavour in the everyday food choices we all make as active participants in the system.

CHAPTER SEVEN

The Challenge of Ripe Fruit

This is a story about a life-altering peach. Never in my entire career has my yearning to find the farmer behind a variety of fruit been so strong. And never was its rediscovery so reminiscent of a detective story. This is the story of the mighty Greta white peach, the same peach with which I started this book.

Back in 2011, we found a variety of white peach that was out of this world, yet as suddenly as she appeared, she disappeared. It created in me a longing that never subsided, even with the coming and going of the seasons. I missed her so and could not erase her memory. Over the years that followed, we tried hard as a team to find her. I wondered at times if we were ever going to be able to enjoy her mesmerising flavour and feel that captivating skin of hers. Like a summer love that is lost forever, I began to think that perhaps her memory was all that would remain, and I should be grateful to have met her. Even her name was a mystery.

The desire to rekindle those precious moments kept the flame of hope alive. We had very little information; all we knew was that those peaches were packed in the Italian region of Campania, where buffalo mozzarella, Neapolitan pizza and some of the best pasta in the world can be found. Hardly a precise coordinate, but this happened to be where Federico and I were heading, on our way to visit some farmers. As we boarded the plane, I saw him scrolling through his phone, cleaning up images to make space. An innocuous activity that set in motion the craziest search we've ever embarked on.

In Search of a Peach

As we landed in Naples on a brisk mid-April morning and headed for the exit, Fede came rushing up to me, visibly excited and in a voice louder than normal said, 'I've found it.' He was holding up a picture, zoomed in to show the side of a peach box, the side with a small white tag that held invaluable information on the origins of the Greta. We had looked for this image for years, unsure if anyone had ever taken a picture of the box. It was destiny that Fede found the image on that flight – nothing short of a miracle. Our well-crafted sourcing trip would undergo some important alterations, as nothing was going to get in the way of finding the person responsible for that tag.

Those tags have the minimum legally required information and represent the organisation or individual who packed the fruit – it could be the farmer, but many times it is not. The product was labelled *Pesche Bianche* (white peaches), and there was a name on it along with an address. To protect his identity, we will stick with his first name, Luigi. What we didn't know at the time was that neither the name nor the town was where we'd find the Greta, but they did get us on the right path.

The address was the obvious place to start, and it proved as useful as a motorcycle ashtray. We arrived in the town and after driving around in circles we realised the address did not exist, so we decided to hit the town square – the older folk there should be a great source of local intel – but we were met with blank stares. Never far from a coffee in Italy, we ordered espressos at the bar and shot them down. It was out of desperation that we turned and asked two men at the end of the counter: 'Does anyone here know Luigi? We source fruits and vegetables and have come down from Milan.' Quick and fast came the reply: 'You're in the wrong town. You want the next town east of here.' Towns here are small, exceptionally so, but this was fate, another stroke of luck as we swiftly left the bar and headed east.

As we approached the next town, we headed for the town square, again in hope someone there would know Luigi and point us to where he lived. More well known in this square, we were sent on our way to Luigi and his brother's place, and through pale brown iron gates we entered a courtyard with a 1980s apartment building rising above. It was under that sand-coloured building that we finally found the brothers. Luigi was sitting on the floor, back resting against a concrete column, hand-grading green asparagus, with a lit cigarette hanging from his lower lip and a grey beanie keeping some heat in. His brother was twice his size, friendly and eager to show us around. Plastic green and yellow crates of just-harvested asparagus were laid all about. No peaches (it wasn't yet the season in any case) and no peach farmers! Naturally, we tasted asparagus, talked produce and as Luigi lit another cigarette, something clicked. 'You guys are looking for Domenico. He's in another town.' This was unreal, a comedy of errors. Luigi hopped on his beat-up scooter, and we followed him out of town; he gestured directions and waved us on our way. That was the last time we saw Luigi and his brother, but I always remember them.

We had to make a couple more stops to refine our directions. Our last stop was at a produce vendor on the side of the road who gave us the last instructions that finally got us to the small market where Domenico had his office. On the 16 April 2014, we met the man behind the peach; three years, one image, one phone call and around seven different people later, we found the Greta, the sexiest peach in the world.

Domenico is one of those farmers who lives for flavour. It is his and his father's flavour crusade that has kept them farming this little-known peach variety. Everything I have said about visual cues of character, the Greta peach has in spades. Intense spotting, a richness of colour I'd never seen in a peach, its skin an intricate tapestry only nature could have conjured. With the cut of a knife, those outside cues gave way to the world beneath, a flesh as magnificent as it was mesmerising. Where the skin was intense in colouration, beneath so was the flesh. And at the bottom of the peach, right opposite to the stem end, the skin was so deeply red it turned a few shades of purple, as if gravity concentrated the colours in the southernmost point. Here, its juice was purple, intense yet with the brilliance only a liquid can display. A lively acidity was the backdrop of knee-trembling flavour. This was a peach on another level. It was on fire – in the good sense of the expression.

The Greta has become iconic at Natoora. Not surprising as it has flavour, character and beauty, encompassing everything you seek in the highest-quality fruit. In Europe, as in the US and Australia, we source peaches from several farmers across different regions. The Greta is the most iconic peach we source and our European summer at Natoora is partially defined by it. Our customers' and our own anticipation cannot be underestimated. It is a work of pure edible art. Finding Greta, and finding Domenico, was merely the beginning. Our shared commitment to delivering a ripe peach has been fundamental to our work. It is a testament to flavour and a testament to the kind of food system we can build if we have a strong vison and conviction.

If I didn't love fruits and vegetables as much as I do, I wouldn't have the patience and impetus to build the complex supply chains that are necessary to bring real flavour back into our lives. Most of us live in urban centres where access to great fruit such as these peaches has been cut off. The challenge is that ripe fruit, a peach that is at perfect maturity, is an extremely fragile product that must be handled with great care throughout its journey from farm to plate. It will bruise easily, any careless movement will damage it, while the right temperature must also be carefully maintained. But finding the perfect peach is possible, and moving it from point A to B with care is also doable. I guess it is down to desire; it is the fuel that enables you to figure out a way to do so at optimum ripeness.

Can you imagine going through all that work we endured to find this monument to nature, and then harvesting it prematurely because you can't be bothered with the challenges and additional work? Isn't the point to get a kick out of life and figure out how to do right by the great white peach? To see the face of a chef when you cut open the peach, perfectly ripe, and expose a Jackson Pollock painting inside?

Instinctively, it makes sense that transporting ripe fruit is harder, more complicated and risky. Nevertheless, we need to dig deeper, as it will help explain why the supermarket model was not designed to move this kind of fruit and level of flavour through the system. The design flaw lies at the heart of the severe deterioration in the quality of fruit readily available to consumers. It will also allow us to explore what the solutions are.

The Fruit Complex

Fruits are more fragile than vegetables, and ripe fruit is the epitome of fragility. Fruit by design is meant to be eaten when fully ripe (in most cases), at a stage when they are soft and full of sugars, that point of utter deliciousness. This is simply not possible for most of us. As we moved from local to more centralised supply chains, we lost ripeness and we lost flavour. It is easy to see why the current system for accessing most of the fresh produce we consume has not been designed to handle ripe fruit. It is a system design issue, just like the bland varieties being farmed today are a product of that same aberration. The supermarket model was not built with flavour at its core. In its design phase, nutrition was irrelevant and farming practices were also someone else's business; the system was built for moving huge amounts of food (and non-food items) cheaply and conveniently. And if the product comes in a box with a year of shelf life, all the better, as it is a lot easier to manage a warehouse full of boxes than one with living product in it, like fruits and vegetables.

The system is not entirely responsible. Consumers played their part in enabling the design of the current supply chains that are incapable of moving ripe fruit, and unwilling to do so. But before we look at consumer demand, it is worth understanding why fruit, as opposed to vegetables, has suffered most from the development of our post-war food system. A good way to do this is to look at a carrot and a peach and follow their journey as they move through Natoora's supply chain.

First, we need to understand what ripe fruit looks like at the beginning of the chain. When fruit leaves the farm, it does not look or taste as it does at the end of the journey, when a chef receives it in the kitchen. If you examine what I would consider a ripe peach as it leaves the farm (as opposed to a ripe peach whose flesh gives if pressed) you will find it hard to the touch. This is slightly counterintuitive, but fruit that wants to be eaten soft should not be harvested at that stage. A few days prior to its perfectly ripe state is when it wants to come off the tree. At this point, you have the optimal balance of ripeness and a piece of fruit with some legs on it. In many cases, fruit harvested at that perfectly ripe stage, if not consumed immediately, will go too far by the time they are eaten. Additionally, it is well known that some stone fruit like apricots and plums benefit from a couple of days of final ripening in a cold environment. At Natoora, when we handle ripe fruit, it undergoes a few days of finishing before it goes on to the final customer; however, even if the peach is still hard, we are still dealing with extremely ripe fruit compared to what the industry moves through normal supply chains.

Carrots and peaches leave the farm in their respective boxes, stacked up into tightly secured columns ready for transporting in a refrigerated lorry. When the produce arrives at our warehouse it is checked for quality, including size, quantity and ripeness. Carrots can be checked extremely quickly: you verify quantity, freshness and sizing in a shipment of carrots in seconds. Carrots cannot be further ripened; the concept of ripeness does not exist with a carrot. Conversely, peaches can and usually must undergo a few days of ripening after receiving them, so at this stage we need to check for ripeness. This requires a team of individuals that can gauge varying levels of maturity in the fruit and make a call as to whether that fruit needs further ripening or is ready to be sold. I want to be clear: you can harvest a carrot before it is ready, but the concept of ripeness is materially different to a piece of fruit. These decisions vary by customer, of course, depending on how quickly they will use it. We can and should deliver a peach in perfect ripeness to a professional kitchen, but when it goes to a supermarket and then into a home kitchen, the extra time that takes needs to be considered. Once quality checks are completed, products are moved into the right warehouse location.

As orders come in, we send the same Sandy carrots to all our customers, be they a restaurant, our own stores or a supermarket. Sizing is the only preference a restaurant might request on a Sandy carrot. However, when we send out a peach, different customers have different requirements of both size and ripeness. Restaurants want perfectly ripe fruit, whereas supermarkets need fruit that has a few days of extra life in them. Furthermore, within our restaurant customer base, each customer can have different ripeness and sizing preferences. These customer-specific requirements add complexity to the supply chain that is internal to Natoora. Complexity adds time, cost and increases the potential for waste. The carrot bears none of these complications.

Supply chains come alive as orders are placed. The difficulty of forecasting order quantities has a significant impact on how supply chains are built and how ripe the fruit they move is. How many carrots need to be quality checked, for example, is a direct result of how many have been ordered from the respective farms, and no matter how good a system you have, you will never buy the exact quantity of carrots you need. Sometimes you've bought too much and other times not enough. With a sturdy product like a carrot, you can buy more to ensure you're never out of stock, because once it is out of the ground it does not ripen and, if stored correctly, retains its freshness for weeks if not months.

If you do the same with peaches, you will throw a lot of peaches away. It is hard enough to try and forecast quantities several days in advance – the lag between purchase order and product arriving to the warehouse – but when you add to that the need to ripen the product for different customer needs, this complexity and variability makes the management of stock levels particularly hard. As a buyer, you rarely make mistakes on carrots, but you will always make mistakes on peaches. To cap it off, peaches are four times more expensive than carrots.

As the product sits in the warehouse and evolves, it needs managing. All product does. You need eyes on it, it needs to move to colder or warmer locations, at times it needs selecting. Selecting is the process of cleaning the product – getting rid of those bad peaches so they don't infect all the peaches in the box. Carrots rarely need selecting, and if they do it is an extremely easy and quick process, whereas peaches need a lot of careful, patient selecting.

In today's food system, we harvest peaches so they behave like a carrot because we have one blunt, inflexible supply chain. Fruit has suffered the most from the industrialisation of our food system and the advent of the supermarket model as a means of distributing fresh fruits and vegetables. The loss of joy and flavour is more palpable, starker, when it comes to biting into what should be a magnificently fragrant peach. Tomatoes are a shadow of what they used to be. The higher sugar content of fruit and their appearance during the heat of summer sets them apart from vegetables, and therefore the gap in flavour from average to great becomes vast and instantly noticeable to the palate.

The difficulty of building a supply chain that moves ripe fruit at greater cost, added complexity, higher investment in knowledge and greater risk of financial loss is a strong disincentive to do so. If consumers don't demand riper, better-tasting fruit, there is no incentive at all. These realities shaped an industry from top to bottom to focus on many other characteristics to the detriment of flavour. Stone fruits today are picked green and ripened off the plant. The state of Georgia, known as the peach state, is a wonderful example of this. A quick search online will bring up numerous websites of Georgia peach farms with proud families standing in front of their harvest in the orchard. The one problem is that they're standing in front of bins that hold up to one thousand kilograms of peaches full to the brim – how ripe do you think those peaches are given they're all still intact, even those at the bottom of the bin? Supermarket tomatoes in the US are almost all harvested green and artificially ripened on the journey to the store. In Europe it is no better, with terrible varieties of tomatoes being farmed because they are sturdier, less fragile, and therefore easier to transport on long supply chains. Flavour, which let's not forget is a signature for nutritional density and healthy soils, has been put at the bottom of the wish list of product attributes. Placing flavour at the bottom of the list is the equivalent of airlines putting safety at the bottom of theirs. Ditch flavour; let's focus on producing cheap, unhealthy food, and disregard the impact it has on the natural world. It's wild.

This is why you don't see a lot of good fruit around. Great tomatoes are extremely hard to find around the world. Mangoes available in supermarkets border on the inedible because price drives their sourcing, and seed companies continue to invest in new mango varieties like the Calypso suited to 'innovative' super-dense orchards capable of producing record amounts of tonnage per acre.

We are not farming varieties that are good enough because of the supply chain constraints and a lack of interest on behalf of the industry. Building flavour-first supply chains takes more investment and conviction, they are harder to manage and carry increased risk and very few are interested in taking on this challenge. There is more nuance to supply chain design – there has to be a human element that is crucial to the system's ability to move exceptional fruit in peak ripeness.

Humanity's Role in the Supply Chain

Over ten years ago we walked into Fondi market, which sits between Naples and Rome, for the first time. An abundance of produce passes through it, and it acts as a processing and distribution hub for the rest of Italy as well as a thriving export business. It is one of the few markets, and likely the largest of its kind anywhere, that still sees small growers delivering produce they have just harvested. The connection with the growers here is special and unique.

It was a difficult journey for us to open up this wonderful market. Fondi moves volume, and at the time we could barely make up a whole pallet, the minimum amount you need to ship produce across cities. Markets are interesting beasts, impenetrable ecosystems with a life all of their own.

We met Domenico Magliozzi on our second day there. A sunny summer day in late June, the air was warm and dry with a damp edge. Vast quantities of industrially farmed produce sat alongside superb, earthy fruits and vegetables. I'll never forget how over a pallet of vine tomatoes we spoke to Domenico for the first time. An imposing figure with vast knowledge, bright green eyes and the pummelling grip of a bear. His love of seafood and cooking is as great as his love of integrity. He commands respect for all the right reasons – reasons that made him believe in us when no one else would. He's a true gem.

Natoora's supply chain is built with hundreds of links, or segments. A great analogy is the body's circulatory system, which is made up of a heart, arteries, veins and capillaries. If we look at the system in reverse, the capillaries allow us to go deep into very specific regions for a single product and the veins are the logistic channels we build to transport the product into the main central vein. At this level, we consolidate and quality check before continuing along the main vein and reaching the heart, which is our central warehouses, that in turn processes and redistributes our produce to all our clients via the arteries – our supply routes. Plotted on a map, our supply chain is far more capillary and complex than anything we've seen in our industry. If we look at a volume to complexity ratio, certain products move in unprecedentedly low quantities within our system.

Domenico heads up our buying in Fondi, one of the largest and most important veins in our system. It is unusual for suppliers to have this kind of presence within a sourcing region, yet we learned early on the value of having our people on the ground. Domenico's presence in the fields has unearthed an old variety of Puntarelle (a chicory variety typical of the area) that is far sweeter and has more internal leaves than what is commercially grown. The farmer, Damiano, understands Natoora's needs, and Domenico works closely with him and our buying teams across all our regions to ensure we deliver the best Puntarelle to our customers. This level of personalisation, the human connection across the network of farms and markets, is what allows us to preserve the attention to detail, the care and the direct relationships that underpin our ability to source the best produce.

Many years ago, I began documenting our sourcing adventures, giving me a broader perspective on our work. It raised interesting questions. What was it that underpinned flavour? Why was one supply chain better at moving ripe fruit versus another? And as I wrote about Domenico and Fondi market, I realised he was far more critical to moving flavour through the chain than any individual product we buy from the region. It was a big moment for me, as I began to understand the answer to those questions. People. The humanity in the supply chain is what made it real. Humanity underpins the quality of the fruits and vegetables we source. Flavour excellence was a result of the individuals in it, their alignment and commitment to a common goal. Farmers, buyers, customers, all obsessed with flavour.

What I have learned over the last two decades is that flavour needs to be designed into everything you do. Every decision in the supply chain must prioritise flavour. Our ability to keep pushing the boundaries of flavour is down to the sense of community we are creating within the chain. How we transmit Natoora's values is fundamental to the community. How we engage with farming partners, the decisions we take with them, and not impose on them, the time we devote to fostering quality, the messages we send with the way we speak to them and what we do and don't ask of them. Our behaviour when things go wrong is critical, particularly when it comes to building trust, if we are all going to accept the risks of moving beautifully ripe fruit.

How do we overcome the constraints of our current food system so we can bring back flavour? Our own commitment has been a driving force in pushing farmers to lean in to ripeness, and the strength of our system comes from the collective force of everyone along the chain. Fewer, larger suppliers are easier to manage. But look at the beauty of working with multiple strawberry farmers, each specialising in a different variety, allowing us to move with the passing of the season. Each farmer playing an important role in the strawberry chain, if only for a few weeks of it. This is the complexity we cherish, the result of flavour-first decisions. Our food system is plagued with financial-first decisions that remove complexity, remove ripe fruit and ultimately remove the flavour from it.

The supply chain must have flex built into it so, rather than imposing your needs, you can be receptive and reactive to the needs of the farmer. Be willing to place orders earlier if it helps the farm deliver higher-quality product, even if it makes ordering the right quantities harder and more prone to costly mistakes. Have multiple suppliers for key lines. This eases the pressure you place on a farmer to deliver product that is not right and adds resilience to the system. We saw this during Covid when our supply chain outperformed that of the supermarkets.

Lean in to the seasons and source a wide range of produce from within a category. Don't just buy green and white asparagus, source five or six varieties from multiple regions. This is another way of building flexibility and resilience into the supply chain while giving customers options to support biodiversity. Let flavour decide how the supply chain must behave and have the flex to allow it to act accordingly. Focus on the right varieties as they can lower the risk for everyone involved.

By working closely with farmers, you can provide security and build trust. Where you can, have a growing plan with your partners as it strengthens the commitment, and the more commitment the easier it is to deal with issues that will inevitably arise. Build a customer base that craves flavour. Build a customer base that gets pissed off if you don't deliver great flavour. They will confer you the right, the ability and the confidence to buy the best produce you can find – and then by working with everyone in the chain make it even better. Invest in education as it will give you a wider customer base, who will push you even further.

Demand is a powerful force that will drive real change at scale. What does this real change mean? A mass adoption of taste and quality across the system. I believe the lack of sufficient demand is the primary reason distributors find it very hard to source amazing fruit. And in a way, most distributors themselves have lost the benchmark of quality, thinking the fruit they distribute is already of extreme quality. That peach is what can push demand in the direction we need it to. That first bite of a peach will change your life and awaken the idea that incredible fruit is a work of art.

CHAPTER EIGHT

The True Cost of Flavour

Changes to the food landscape since the 1950s have led to serious consequences for the food system, our health and that of our soils and ecosystems. Consumer demand guides the system, with our food choices acting as a navigator, telling the system the direction it should take.

Nothing speaks louder than data on the disposable income spent on food over the last decades. A study by the US Department of Agriculture (USDA) found that the amount of disposable income spent on food at home in the US declined from 23.9 per cent in 1929 to 10.7 per cent in 1997.[1] The report says the decline is a sign of a more developed nation, which I find laughable. Surely some of the decline is inevitable as incomes increase over time and less cooking takes place at home, yet the data shows a shift in what we value, in what we choose to spend our money on. A second report by the USDA that looked at data spanning nearly 60 years from 1960 shows a decline from 17 per cent in the beginning of that decade to 9.9 per cent in 2000, with the amount of disposable income spent on food staying steady into 2019 at just below 10 per cent.[2] This data confirms the cheap food trend the supermarkets instigated and, in turn, the declining value we apportion to good food. Not everyone has the luxury to spend more of their income on food, but many of us do, enough of us to make a difference if we were to value it correctly. The more joy we can find in food, the more money we will divert to it.

As the Second World War ended, the industrialisation of our food system picked up pace. Household dynamics changed, and consumer behaviour shifted, paving the way for the supermarket concept to rise and become the preferred place for people to buy food. But it was not simply a more convenient and affordable place for families to buy apples, milk and bread; these new stores were the home of processed and ultra-processed foods. The rise of the supermarket and the rise of highly processed, artificially flavoured foods are one and the same. Something similar happened with tobacco where, in the early days of the industry, people were unaware of the severe health concerns until it was too late. Shoppers entering a supermarket in the 1970s or 80s were fooled into believing these shiny new products were real food. As demand for Kraft Mac and Cheese went up, demand for real cheese went down and the dairies transforming fresh milk into nutritious cheese were forced to make less of it, while Kraft factories doubled their shifts.

The Dangerous Decline in the Value of Food

Ultra-processed foods are based on a concept revealed by American researcher Howard Moskowitz, which he coined the 'bliss point'. Moskowitz was hired by the US Army to try and solve a serious problem with soldiers out on duty; the soldiers were lacking the necessary caloric intake because they were not finishing their MREs (Meals Ready to Eat), which were designed to be shipped and stored easily so they could be sent anywhere around the world. What Moskowitz discovered was that the brain initially likes food with bold flavour but over time it gets bored of it, overwhelmed, as it seeks diversity. He also found that soldiers did not bore of bland foods and, while they were not as enthused by them, they could eat much more without feeling satiated. This discovery set the foundations for ultra-processed foods by combining food that excites the brain and tastebuds but never to the point where you are satiated, and therefore you continue eating them. We have all experienced this with a bag of potato crisps – as the marketing brains at Pringles say: 'Once you pop, you can't stop.' The bliss point is the precise combination of sugar, salt and fat that achieves this.

Worryingly, the only food that naturally contains sugar, salt and fat is a mother's breast milk (it is the only food we consume at that stage of our lives, so it has to be a complete food with everything we need for healthy development of body and mind). I say worryingly because naturally we are designed to want that combination of protein, sugars and fats, and therefore it is easily taken advantage of by today's food industry – it is no wonder the ratio of those ingredients is very similar to a lot of the junk food we consume. We engineer food that tastes good to us, has zero nutritional value and never fills us up, and our consumption of these foods has been on the rise ever since. The data is astonishing. In England, per capita sugar consumption (including sweeteners) went from a mere 1.8 kilograms per year in 1700 to 45 kilograms in 1950.[3] In the US, yearly consumption has grown from 1.8 kilograms per year in the 1700s to 61 kilograms in 1989.[4] (These numbers are based on how much food disappeared from the food supply and don't account for wastage, so actual intake per capita will be somewhat lower.) While per capita sugar consumption has started to come down from a peak in the 1990s, we continue to eat way too much of it.[5] According to the

American Heart Association, adults, young adults and children in the US consume two to three times the recommended amount of added sugar (sugar or sweeteners not naturally present in fruit, for example, but rather added as an ingredient). Our bodies' inbuilt systems have been hijacked for pure profit, with the sole intention to get us to eat as much as possible.

These changes have had serious, long-lasting consequences. The worst of them is the decline in value consumers apportion to fresh food. Our unwillingness to recognise the true value of an incredible melon makes it impossible in economic terms to sustain the farming and processing of flavour-first food. When we attribute less value to a well-farmed melon that tastes amazing, we are less willing to spend money on it. You don't need an economics degree to see how the choices consumers make get in the way of a farmer's ability to produce a great fruit. If our children prefer Froot Loops cereal to real fruit, we have a problem.

Our food choices are driven by three forces: what we desire, what we can afford and what we have access to. All three are shaped by what we choose to pick off the shelves, the demand we stimulate. Interestingly, and perhaps less evident at first sight, is that all three forces can be influenced through flavour.

Desire

Our food system needs rebalancing and so do our consumption habits, our lifestyles and the choices that define them. To do so, we need consumers to want to spend more money on food. We must create the desire. I believe flavour can unlock the desire that gets people to pay more for a better-tasting apricot. Value underpins conscious consumption because consumption becomes sustainable when there is a positive financial outcome for everyone involved.

In agricultural terms, increasing the price of a cabbage is what allows the farmer to focus on flavour, with positive ramifications for a nutritionally superior cabbage, improved soil health, increased biodiversity and greater joy. Currently our industrial agricultural system excludes the damage to soil and local ecosystems – the externalities, as they are known – from the price of the product. When you buy that absurdly cheap, flavourless cabbage, someone other than the farming enterprise that grew it will pay for remediating the damage.

Making the situation worse is that bad food is cheap. Made in large factories, ultra-processed foods are chemical marvels that use the cheapest ingredients and intense processing methods. The consumer thinks food should be inexpensive. If a Dorito makes my tastebuds go crazy, hitting that bliss point, while the tasteless apple bred to grow in a climate not adapted to it and stored for eight months is more expensive, go figure what happens to our food going forward.

Markets will price to what consumers can bear. We don't value good food anymore. We don't understand what it takes to grow and produce good food, nor do we appreciate the consequences of our choices. All that cheap food has a cost to our health, our communities and the planet.

Europe's horsemeat scandal of 2013 serves as a sad reflection of this reality. Testing found many of the beefburgers sold in supermarkets contained some horsemeat, and in some cases pork, rather than being 100 per cent beef as advertised. There was uproar, but what no one commented on was the cost of 'value burgers', the category supermarkets classify them under. Consumers should have been thankful it was horsemeat, a perfectly acceptable source of protein, and not something else. What goes through our heads when we pick up a packet of burgers or sausages that cost less than a cup of coffee? At the time of writing this, you can buy a whole chicken in Asda, one of the leading UK supermarkets, for £2.64 per kilogram; that's £3.96 for a 1.5 kilogram bird. Asda will likely make a 30 to 35 per cent margin, meaning it has to buy the chicken for around £2.50 – that's £1.67 per kilogram. Someone must hatch, feed, slaughter, clean, package and deliver the chicken, and then add the marketing required by the supermarkets to list the product. If the producer needs to make 30 per cent on that chicken to cover overheads and make a profit, it equals a production cost of £1.75 per bird. No one should want to eat a chicken that costs less than a bus ticket to produce.

As we revalue food, it changes what we can afford. The willingness to pay more for a good product goes up. Of course, there are always limits based on income and the different expenses families have; however, affordability is also based on where we apportion value to in our lives. If you love basketball, you will gladly spend money on tickets to watch a game over spending it on a meal out, whereas someone else will choose the meal. The amount of disposable income we spend on food is down to us: it is a personal choice we control. I'm more than aware not everyone is as fortunate. Many individuals rely on the help of the government or non-profit organisations to feed themselves, to find much-needed nourishment. They don't have a choice, they don't have the luxury of these decisions, but those of us who do have a responsibility. We have a responsibility to those less fortunate than us, as the food system we choose will be the same one they are fed by, only they don't have the power to change it.

The more joy I get out of eating an amazing artichoke the more I will value it, and it will then have more relevance in my life and the lives of those around me. And if I can see the good that has come out of growing that artichoke and the health benefits for those who eat it, I should value it even more. When value increases and desire to spend increases, we take some important steps forward in fixing our food system.

Affordability

The issue of affordability must be tackled from two opposing sides. We've addressed the need to be willing to spend more on food, but we also need the price of quality, nutritious food that is full of flavour to come down. Both are necessary to make food more affordable, essentially narrowing the gap between the two opposing forces.

Perhaps counterintuitively, scale is the way to tackle price affordability. Scale is a very valuable tool. Often it carries negative connotations, understandably, given how many 'big' things we can point to that have lost soul, direction, purpose and integrity. Industrialisation has been all about scale – how do we get bigger and go faster. In the same way we utilise the capitalist model to drive change, we can use scale and its benefits for good outcomes. A great outcome is farming beautiful vegetables at scale and driving prices down by virtue of those economies of scale. The larger the volume of product you move, the lower the cost of transport and the lower the price. Using processes allows for scale, which in turn provide cost benefits. Additionally, mechanisation, when it is applied with care so as to not degrade the quality of the product, can lead to efficiencies that help lower costs and drive prices down.

In our London mill and bakery, Alma, we have set out to prove that we can source fantastic grains and stone mill them in-house, retaining all their nutrients and then turning that freshly ground flour into loaves full of nuance and flavour that are highly nutritious. We're confident we are baking some of the best bread in the world. But on its own it is not enough. Alma was co-founded with Graison Gill, a friend and one of the most talented bakers out there. Over the years, I've had many conversations with Graison on scale and impact. As our conversations on building a mill and bakery in London strengthened, the spirit was crystal clear: build a bakery capable of making bread of impeccable integrity while having meaningful impact. To do this, we built Alma with the rigour, processes and equipment for large-scale impact. Thousands of daily loaves will enable Alma to make a difference to the wheat fields and, in turn, mean tens of thousands of consumers are experiencing bread the way it should be. Designing scale into the model early on is what allows us to produce outstandingly nutritious bread at a price that is affordable to the consumer.

As demand for flavour increases in the system, costs of production and transport can come down, making quality food more affordable. It is all about balance. The answer is not mega-farms, but also not micro-farms. I'm not arguing for them not to exist – all farms should form part of an efficient food system whose output is flavour-first produce. On the large scale, those mega-farms will need to be phased out over time – at the moment they are too embedded into the global farming system to easily unplug them from one day to the next. Micro-farms, on the other hand, should thrive, and as a supply chain gains scale they can feed their smaller harvests into a well-oiled and efficient system.

While we focus on the special seasonal fruits and vegetables, those ingredients that are the star of a dish (in the US they call them centre-of-plate), the realities of both our food system and that of our customers means we do provide a wide range of fresh produce. Even high-end restaurants need to balance their finances and they too buy industrially farmed onions and carrots for making stock. At Natoora we have been looking at ways to tackle this by utilising scale. We have a programme called Better Staples, utilising volume to improve the quality of the basic products we source, which tend to come from industrial agriculture. Many years ago, Fede and I met Girolamo and his late father, Carlo, who for the last ten years has been our primary source for all the flat-leaf parsley we buy in Europe. That meeting and the parsley overture that ensued was not only eye opening for us but also sparked the idea for Better Staples. Girolamo is one of those beacons of the farming future: young, insanely talented, knowledgeable and forward thinking as he focuses on flavour and the growing of healthy, safe food.

Girolamo delivered an eye-opening lecture on current parsley production. He went into a great level of detail on the intense, chemically fuelled production methods. Enlightening, and scary at the same time, it gave us the impetus we needed to make a change. What we learned was that most of the flat-leaf parsley that restaurants use comes from highly industrialised farms, heavy on chemical inputs and growth hormones in a quest for yield and affordability. Hormones are applied to the plant at an early stage, to artificially stimulate faster growth, like using steroids to bulk up muscle mass. The result of this intensive farming is thick, long stalks with large, tough leaves, adding mass to a product sold by weight. It also shortens the time from seed to harvest, driving costs down.

In the early days we sourced beautiful, delicate, thin parsley from a handful of small growers. Unfortunately, as professional kitchens in London had become accustomed to this beefed-up yet flavourless parsley, Natoora was steered in the wrong direction. The funny thing is, industrial parsley is an aberration in the kitchen. Roughly chopped, it is unpleasant to eat and adds little if any flavour. Leaves must be picked from the stalks as they are useless, usually hollow and always dry. Parsley should be tender, leaves, bizarrely, should be sort of rubbery to the touch as if coated with the finest latex, while stalks should be moist and a delight to eat. The funny thing is that while a kitchen will see the massive bunch of parsley arriving, by the time you've picked the leaves from the stalks you've wasted more than half the weight, and then you lose the rest to a deficit of flavour – a very tangible false economy. What seemed affordable is in fact costly, and not just to the restaurant finances but to our health, that of the farm workers and our soils.

So noxious is the farming of this parsley, when you push hard with chemical fertilisers and hormones, that you damage the soil, and furthermore farmers aggressively push up land yields by cutting down the time between harvesting and replanting a tunnel. A very strong chemical gas called Vapam, technically a liquid that vaporises, is sprayed on the land post-harvest to kill off all weeds in preparation for replanting. Once you spray you close off the tunnel for two weeks, or thereabouts, and leave it to complete its scorched-earth mission. Workers are known to enter those tunnels prior to two weeks, when dangerous gases are still working to push yields up even further. This creates the well-known cycle of deteriorating land and ever-increasing aggressive use of chemical fertilisers and pest controls to deliver a product that is potentially dangerous to human consumption. And, as nature would have it, the resulting product is nutritionally inferior, lacks flavour and has physical qualities that make it undesirable in comparison to the real thing. Fede and I regularly spoke about the 'good parsley' and what a shame it was we could not source it due to customer demand. Girolamo, with his eye-opening lecture, convinced us there was a way.

Since then, our basic parsley has beautiful, tender, light, deep-green leaves with ultra-thin stalks that are a delight to eat, vibrant and juicy. To achieve this, Girolamo focuses on sourcing parsley seed that is slower growing and more flavour forward, with no propensity for its stalk to hollow out and toughen. He then follows this selection with a farming regime that is focused on soil health, natural inputs like inoculants to stimulate microbial life and using low till to reduce soil disturbance. Girolamo also constantly explores the latest non-invasive pest-control techniques, which ultimately result in healthier soil that delivers greater nutrition to the plant, and flavour. No one asked us to do this, and no one would have stopped buying our parsley. We did not need to invest sourcing resources to switching parsley, but we did. Utilising our collective purchasing power to revert into the supply chain, we scaled to source a product whose farming practices are far better, with full transparency, at a comparable price point, thus enabling the shift. Mass affordability of fruits and vegetables will need to come via some of this supply-chain-design thinking. Scale has an important role to play in a system that is so vast.

Companies like Natoora need to build those supply chains that are flexible, complex and diverse and can absorb product from very small farms. But without a large swathe of what I call 'accessible scale' farms, I don't believe change can happen. These farms sit in the large space in between micro- and mega-farms. Similarly, Alma's ambition is to develop into an accessible scale bakery.

Access

Accessibility might be the most challenging force of the three. What we desire and what we are willing and able to pay are more straightforward. Accessibility has a lot to do with geography, existing infrastructure, even racial and demographic considerations, particularly in the US. The concept of a 'food desert' (populated areas where fresh food is inaccessible) is very real and terribly sad. I do find, however, that having access to fresh, healthy food is also a function of consumer demand. We should never forget that the moment we stop buying something as a group, it won't take long for that item to no longer be on the shelf and, over time, no longer be produced. Look at what happened with the 'New Coke' that appeared in 1985. Due to poor consumer demand, it became extinct, and there are countless examples of products like it that are no longer gracing supermarket aisles. What we demand is a sure way of addressing access to good food.

Some of these consumer-driven changes are evident in cities as neighbourhoods undergo gentrification. No doubt income plays a major role in food choices; many would love to buy organic but can't afford it. But if change can start with those more fortunate then so it should. What you can see through gentrification is the change in product range at a corner shop. Organic choices begin to emerge, better quality and more expensive cheeses, a wider selection of fresh product. As consumers move into a neighbourhood and shift demand, local stores follow and meet that demand. It is a clear indication of the power of demand, and we should exploit it to make good food more accessible.

Having spent close to two decades building fruit and vegetable supply chains across different regions around the world, I have gathered significant perspective on how we move fresh product from farm to professional kitchens and consumers' homes. Culture is a leading indicator of evolved supply chains – by evolved I mean supple, reactive and decentralised with important nodes, central hubs in major cities that absorb and redistribute large volumes of varying qualities of product. It makes the movement of fresh produce very cost-effective; efficient, fast, capable of moving high-quality produce almost effortlessly. You find these supply chains in Europe and some parts of Asia, and the wholesale markets play a fundamental service as they are the circulatory system's equivalent of a major vein.

The reason that these more evolved supply chains exist in certain regions is down to what customers demand. The connection between a nation's culinary choices and the supply chain that emerges to serve it could not be stronger. In Tokyo's produce market I had one of the most memorable meals of my life after a tour of the market. At the door signposted by the customary linen banner, we were greeted by an old lady with brown leather sandals over white socks, a pocketed apron over her shirt and neatly tied burgundy headscarf holding all her hair back and fully exposing her beautiful face. Hosted by some market locals, we were taken upstairs and, after some vivid exchanges with the old lady, food began to arrive. First some pickles, followed by a bowl of raw tuna, then a light sauce and by the end each one of us had an incredible variety of insanely delicious food – pickled, fermented, some raw and others cooked. This is what market traders come and eat every day. In stark contrast, in US or UK wholesale markets the only food available is ultra processed.

In France, moving artisan-made, high-quality fresh cheese is extremely easy. Most households eat a cheese course as part of a meal, which is why when you open the fridge of any household in France, you're likely to find at least three or four types of cheese in there – and I'm not talking about block cheddar. These are important correlations that speak to consumer demand – value and supply chains evolved to deliver what people want to eat. Trust me, you will not find one single wholesale market in the entire US that sells artisan cheese. The same goes for the UK. This is why consumer demand is the tip of the spear.

The importance of these nodes is twofold: a logistics infrastructure that underpins them, which as I've said before is very capillary, like our circulatory system, that can draw in vast quantities of produce of varying quality and efficiently find a home for it. You won't find produce from small farms at Hunt's Point, New York's wholesale market, which is one of the largest in the world, but you will in Paris's Rungis market or Milan's central market. Logistically, if Natoora wants to buy produce from a small farm in the north of France or southern Italy, it is cheap and easy. This is because of these central markets that are built to absorb even small quantities of produce very efficiently. It is easier and less costly to move a pallet of fresh vegetables from Sicily to London than from Cornwall; this is a direct result of more evolved logistics networks and food systems geared for flavour.

If we want to build a supply chain and create a market, as we've done for flavour-first fruits and vegetables, it needs to be rooted in sound and market-approved economic metrics. Everyone along the chain must benefit, and everyone along the chain must understand the need for efficiency to open up the market to significant change. Pricing is fundamental to this exercise, and one of the most misunderstood pieces of market strategy. Before we attempt to build a sexy supply chain, we must understand the costs of doing so and how to build in the pricing at each point that allows the system to be self-sustaining. If I charge customers too little it simply won't work long term. If I want more customers to be on board, I need to find efficiencies so I can drive costs down, and yes, some of that means negotiating prices with farmers and producers.

If a new supply chain is not financially sustainable it will be built on artificial foundations and have zero positive impact in the long term. Ultimately, if the economic model does not sustain itself and everyone in the system, sustainability is a pipe dream. You can invest in scale, you can subsidise early on, but you must build financial rigour into the model. We did this to great effect with Good Earth Growers (GEG), Sean's farming business in Cornwall, which we supported from its early days. Natoora committed to divert purchases of some key products like green kale, which we were buying in markets or that came from conventional farms we had no contact with, to them and in doing so guarantee consistent volumes. Our commitment allowed the farm to scale, removing most of the associated risk. Then, by subsidising the cost of transport and margins – effectively pricing the produce at what would be long-term unsustainable margins for Natoora – we created a market for flavour-first produce. Over time, were able to claw back margin to sustainable levels by lowering transport costs and increasing some of the pricing as the products, with their higher ethical credentials and exceptional flavour, took hold.

I have seen too many food businesses seeking to change the food system with business models that make no economic sense, and none have survived. Many of the farmers who for a time were buoyed by the higher-than-market prices paid for their produce by these food start-ups, along with super-prompt payments financed by venture capital money, have lost out in the end. These farms end up selling a big portion of their produce to one customer who is sustained by aggressive capital, and when capital dries up, the farms are left with produce that won't fetch the prices they have been used to, creating a serious problem. There was a well-known company in the UK called Farmdrop that is a startling case study of what I'm describing. Centred on the very laudable idea that the current food distribution system is not working, they sought like many others to build a new system by connecting consumers directly with farmers and producers. Heavy in their marketing was the fair prices they paid to their suppliers – how they gave a larger percentage of the pie to them.

They garnered a lot more attention than they deserved, for at no point in the close to ten years they were around did they have anything like a viable model. The intentions were there, but the foundations were artificial. What good is it selling a fantastic chicken for £10 to a thousand customers a week if it should cost £13? Looking at Farmdrop's published accounts, in 2019, the year before they declared bankruptcy, their pre-tax losses were £11 million on £5.4 million of revenues, and in 2020 on revenues of £11.8 million they posted a pre-tax loss of £10 million. Even on the back of a doubling of revenues thanks to the Covid pandemic, Farmdrop was barely able to generate any headway into profitability. Going from losing 2x revenues to 1x revenues is hardly a sustainable model financially, and neither can we classify it as fostering sustainability. My experience tells me the main issue was pricing.

Oscar Zerbinati, who farms those surreal pumpkins in Mantova, northern Italy, must face a simple reality that by removing up to 25 per cent of the water content during a post-harvest curing process that results in life-altering flavour, he must compensate that loss in yield with a significant increase in price. Flavour is how we can justify it; it is the only way. There is also additional work involved in curing those pumpkins and far greater risk of crop loss, as mould can occur during the curing process. Oscar has in the past lost a large percentage of a harvest due to this mould issue. He must charge more, and by fighting for flavour he can.

The whole system from farmer to consumer needs to value and integrate higher production costs, the true costs of flavour, into the pricing. By doing so it forces a sound business model that has legs, that is real and will survive. Its survival is what sustainability is all about, and that is the first step in the journey. Once that is in place one can think of all the greater good one can accomplish, especially by utilising healthy profits to stimulate further change.

New models in the food system need to build supply chains that are designed to drive value across the whole chain. At each point the right price needs to be paid. It is only by building real demand that we can think about real, lasting change. By real demand I mean someone paying the true cost of a product. Back to the Farmdrop example – what is real about saying we support farmers to the tune of £5 million a year in purchases when those sales are artificial? Customers were not paying what it cost Farmdrop to source and deliver those products and therefore that demand is not real. If prices were 25 per cent higher, would they have sold as much food, would they have supported as many farmers? That is the real challenge. Get customers to value the product and pay what it is worth. If over time prices can come down through scale that is a bonus.

If the real cost of producing flavour and all the benefits that come with it are priced into the supply chain and, through education, we can engage consumers to pay that price, we have now created a market for flavour. We can build that market over time because the foundations are sound as we continue to engage with consumers and let flavour provide the joy that stimulates demand back into the supply chain. This is how I've learned to build flavour into the core of the supply chain. I've seen it work miracles, and I've seen models with great intentions unfortunately fail because the true cost of flavour was not reflected into the model. Sustainability must start with financial coherence.

Accessible Scale: The Right Kind of Scale

In Calabria, as we approach down the coast from the north, to the right the first glimpses of vast Mediterranean blue appear. Along this stretch of coast that begins in Falerna Marina and extends down to the village of Tropea and roughly 1.5 kilometres inland, is the only piece of land where you can grow the Cipolla Rossa di Tropea as per the *Indicazione Geografica Protetta* (IGP) regulations. As you'd expect, there is a lot of olive around; green is everywhere with a certain freshness in the air. Terracotta roof tiles seem to pop out of nowhere, beautifully aged, discoloured to the point of perfection as if they were just another plant setting roots on the hill.

Natale is the third generation of the Santa Croce family to farm on this land, having quit university to fully immerse himself in the family business started by his grandfather, who originally was a tenant farmer paying the landowner a percentage of his crop earnings. His father manages the commercial side of the business along with the plant nursery that was set up in 1992, while Natale oversees all the growing.

Santa Croce is a beautifully run family business supporting two hundred and fifty employees across a farming operation and plant nursery. Their output is staggering for a single product of this quality: five thousand tonnes – that is five million kilograms – of onions per year. Ninety per cent of that is sold fresh, which means these onions are harvested, selected, cleaned, trimmed, bunched and packaged within a twelve-hour period. When you can produce for twelve months of the year, while rotating and regenerating your land, and actively managing a seed selection programme plus producing all your own seed, all one thousand kilograms of it, you understand the scale and specialisation of Santa Croce and how far forward they've taken their profession. These are the farms, the agribusinesses of the future, dedicated to scaling with a vision to deliver better quality, better flavour, to more and more consumers around the world. Within this context, it is worth noting that Natale oversees the growing of unquestionably some of the finest onions in the world. This would not be possible without specialisation – this is what accessible scale is all about.

As we approach this new dawn in consumption and our relationship with the planet, we need farming systems geared to producing highly nutritious food in harmony and support of soils and wider ecosystems, but importantly, with the necessary scale. We cannot afford to work on the fringes of the system; in order to see profound, material change, we need to meet consumers in the middle.

We cannot expect to draw consumers to better food by asking them to make wholesale changes to their lifestyle habits. Tapping into existing distribution systems is the only way to achieve this level of change and, for that, we need to produce great food in large-enough quantities, otherwise that food will be relegated to the weekly farmer's market. We need a farming model that can sustain lasting change in our food system.

Specialisation is key. It does a number of things for the farmer, for the quality of our food, and for its ability to scale and impact real change. I call it accessible scale because farms are capable of achieving a scalable production model that allows a fairly limited number of employees to produce, at incredibly high levels of quality, a consistent product at high volumes without the need for the scale of industrial farming. An accessible farming environment that can be reproduced and scaled by nature of its specialisation. I see it as the farming equivalent of *mittelstand*, the German family-owned businesses that represent the core of the country's industrial manufacturing machine. *Mittelstand* companies have achieved great scale while remaining proportionately small in comparison to the industrial giants.

The issue of scale is really critical. As I've mentioned before, change can only happen if we meet the consumer where they shop. With more than 80 per cent of fresh food going through supermarket tills, we must leverage the existing distribution systems but to do so food needs to be produced at scale. As a farmer, or as Natoora, you cannot approach Whole Foods Market, or any supermarket for that matter, with ten boxes a week of heirloom tomatoes. But one highly specialised farm like Santa Croce can work not with one but multiple supermarkets. In Italy, their onions are widely available in many supermarkets around the country.

It has taken the family decades to develop the expertise to breed the multiple varieties that allow it to farm some of the finest onions in the world at scale with considerable attention to the preservation of their land. The multiple varieties is what has enabled them to have a much longer growing season and develop the market for fresh Tropea onions, as opposed to the traditional dried ones that were exported into the US back in the day. It took Natale's father twenty years to accumulate thirty varieties of Tropea onion seed. Talk about biodiversity and resilience; on its own, this is great work for humanity.

Natale spent two years in France learning organic agriculture and responsible farming methodologies for onions. Once back at home, he had to figure out a way to apply what he learned to his climate, his soil and the specific variety, which were all different from what he learned in France. Some of their land is sandy like a beach, while other lots are sandy clay. He layered this experience onto his father's thirty years of growing expertise, mostly producing strawberries which lead to significant soil degradation. This experience marked them, and along with local tradition and a passion for flavour, has led this family to modernise and seek to build a farming business for the future. Natale is no fool and therefore he speaks of the two things he most protects: his soil and his seed, as the longevity of his business relies on them.

Specialisation of this level requires careful land management to protect the soil. The planting programme is so precise that it takes into consideration the time and climate necessary to regenerate the soil, considering he has the same crop being harvested fifty weeks of the year, grown on the same total land area. Natale uses a technique called solarisation during the summer months, which consists of covering the soil with a clear plastic tarp that absorbs and traps the sun's heat, and significantly raises the temperature of the soil, thus killing off weeds and pathogens and avoiding the need for harmful chemicals or other inputs. He also rotates his crops and follows onions with brassicas or barley plantings. Barley is a great resource for reintegrating minerals into the soil. Depending on the year, the weather and timings, Natale will incorporate the barley back into the field for maximum effect. If time is not on his side, he will harvest the barley prematurely as the roots will have done a lot of the work, and ready the land for replanting.

As we stand by a plot full of onions going to flower, the magic of this place fills you. A gentle sea breeze comes in from over the hedge, essential in cooling and maintaining a stable climate. Natural ventilation of the finest kind. The soil is not soil, it is sand, a beach really. You could unfold a lounger, bare feet caressing the warm sand, and you could try to count the number of flowers each onion shoots up out of that central cane. Big masts of candy floss all waving calmly in the breeze; the peace is comforting, it envelops you.

Green canes as hard as bamboo lift the flower head like a trophy, exposing its entire beauty. Circular perfection, a complete 360 degrees, made of miniscule, green-stemmed white flowers, all a perfect yet different length. Not by chance Natale has the same 360-degree vision, that circular perfection, which has allowed him and his family to create a model for the future of agribusiness. An insanely specialised machine, producing world-class food, reliant on both technological innovation and intense appreciation for the value of manual labour to deliver quality. Believe me, this is highly revolutionary. As we stand looking at the future of Natale and his family's business, those tiny black seeds that the flowers will produce will hold inside them the decades of work that came before and all the wonderful future ahead. Think about that for a while. I can assure you there is magic here, magic in abundance.

Most farmers would find it hard to believe that a monoculture farm was producing this level of quality while nurturing the land and farming in an exceedingly responsible manner, with rarely any chemical inputs. The same is true of hundreds of farms around Europe. Farms that benefit from a tailored, very capillary supply network to move quality product quickly and cost-effectively. As these farms are part of a wider supply chain, the scale and potential impact is even greater.

While diversification is a good farming practice that can drive farm security, it prevents a farm from reaching accessible scale. There is no conceivable way that I see us replacing the current distribution model; we need to harness its strength and customer reach to generate the necessary customer-driven demand. This will create a loop that constantly sucks better-quality product into the supply chain and out through to the consumer, ultimately via the supermarkets and other mass-market outlets.

Specialisation is the last piece of the puzzle and arguably the most important, for it is here that farmers really need to take note and believe the change is worth it. Specialisation drives standardisation, bringing cost prices down. It also fosters greater advances in quality and a deeper understanding of the plant and its environment, both critical to mastering flavour.

If flavour is the key that unlocks consumer demand, we need farmers to adopt systems that promote flavour.

Much has been written about the negative effects of specialisation in farming. But, as we have seen, specialisation does not have to mean mega monocultures heavy on inputs detrimental to soil biology. In Michael Pollan's timely book *The Omnivore's Dilemma* he concludes by presenting a holistic farm as a solution. An entirely self-sufficient farm, which, by its nature, sits in a very closed, local community. Not specialised by the very model it subscribes to; it cannot move large quantities of a single product and does not scale. I am confident that specialisation can occur while rebuilding soils and managing pest and disease pressure without the need, or with minimal use, of chemical pesticides. We have seen many farmers we work with thrive under this model.

Winter Moon Roots, based in Massachusetts, is one of them. Interestingly, Michael Docter, who set it up, was one of the pioneers of the Community Supported Agriculture (CSA) programme in the US where farms sell upfront shares in future harvests to the local community. Locals receive a weekly produce box of what's available and therefore farms must grow a wide variety of produce for this model to work. Specialisation is not an option. The two of us have had great conversations on this topic as we see eye to eye on the concept of farm specialisation, soil health and scale. As a pioneer of the CSA model, his perspective has been enlightening and reassuring. He confirmed my hypothesis that reducing diversity at a farm level is beneficial for the farm, for product quality and for the system. We spent time by the river out the back of his house eating tacos while we discussed the inefficiencies of the small-farm model in the US. Winter Moon Roots, now under the stewardship of Rosendo and his wife, Genevieve, who bought the farm from Mike after years working with him, focuses exclusively on winter roots. The result? Some of the finest carrots, turnips, parsnips and beetroot I've ever had anywhere in the world. Produce that would stand out in Europe, and that is hefty praise. More, much more of this is needed in markets like the US if we are to infuse flavour back into the system.

CHAPTER NINE

Education

I am often asked what Natoora's greatest challenge is. Worryingly, my answer is very evident: education. We need to eat to survive, and our bodies are designed to tell us when we're hungry. But consumerism is all about tapping into human psychology because when you do so successfully, you lean in to inbuilt primal responses that consumers can't switch off. By taking advantage of changing consumer habits, supermarkets and processed food manufacturers sabotaged consumer psychology for their own benefit. If we can design food that tastes good and doesn't fill you up, as we've done with potato crisps, then we can sell lots of them. Your body loves the flavour, wants to eat more of them and, because they don't fill you up, your body keeps telling you to eat more. Nestlé and other food manufacturers of its ilk got so good at understanding human psychology through actively studying consumer behaviour that they could manipulate it to their advantage.

Sabotaging Consumer Demand

What I'm going to share is common knowledge in the market research industry – in the departments responsible for understanding consumer behaviour and how consumers react to changes in packaging, size, messaging, etc. These companies realised that the larger the size, the more we consume. Let's say that in a family's weekly visit to the supermarket, it buys two bottles of milk, a case of sparkling water, little yoghurts for kids' lunchboxes, a box of cereal and a bag of potato crisps. That bag of crisps, irrespective of size, is eaten and replenished every week. By making the bag larger and incentivising consumer behaviour through price, they increased the volume of potato crisps consumed. The same has happened with Coca-Cola, Pepsi and all junk food, which now comes in larger packages. When I was growing up, after football practice with my friends, four of us would share a one-litre, ice-cold glass bottle of Coca-Cola at the local kiosk. Today, most people buy a half-litre plastic bottle and finish it off themselves.

I'll never forget a French family who moved to New York in the late 1990s and their reaction to the gallon-size bottles of orange juice and children's yoghurts that were twice the size of what they found in France. In the US, fridge doors are designed to fit these four-litre monsters. They're even marketed as a feature. Food manufacturers and supermarkets together tricked families with feelings of value and convenience, yet what they really wanted was for them to consume more of this unhealthy, flavour-engineered food, which is cheap to produce and easy to transport because of its industrial nature. Cheap ingredients, highly processed. Why bother with figuring out a way to move beautiful ripe nectarines when it is far less complicated and more profitable to move bags of Doritos? Blindfolded, we lost contact with seasonality, we lost contact with flavour, and over time lost a most precious resource: knowledge. We stopped cooking and gave up the ability to feed ourselves a wide, biodiverse diet full of interesting tastes and micronutrients. We lost the ability to make the right food choices because we live within a food system full of fake food with fake flavour. Our bodies think Doritos are good for us and that strawberries should be available year-round. This is a tragedy. If, as a society, you cannot make the right choices, and you buy strawberries when you shouldn't, the food system goes haywire, as it indeed has.

What and how we want to eat is the consumer's responsibility, but customer preferences changed. This de-education was fuelled by the industrialisation of the food system for sure, but consumers also recalibrated their appreciation for quality, wanting fruit not to degrade as quickly and to have vegetables that are easy to prepare and always available. In our Chiswick store one day, I heard a mother walk in asking for broccoli and when she was told it was not in season, perplexed, she replied: 'Well, what is my child going to eat?' Or, God forbid, an orange goes mouldy in the fruit bowl inside a room that is constantly 20°C. Even in highly food-educated cultures like Italy, there have been visible changes to consumer behaviour as de-education has taken hold.

In a democratic society, consumers hold the ultimate power, but we cannot forget that education is an imperative ingredient. Without it, we fall prey to misinformation, marketing, false aspirations and ultimately the manipulation of intrinsic human quirks. We have recent evidence that companies like Facebook and TikTok have manipulated human psychology, our deep-rooted evolutionary responses and behaviours, to create an addiction that fosters excessive consumption of things we don't need.[1] Knowledge is the only way we can counter these forces, and we acquire knowledge through education.

Education can reinforce our desire to eat a healthy diet. What our tastebuds appreciate and respond positively to is formed by education. As we expose ourselves to a wider variety of foods, different flavours, we expand our palate and evolve it. No one is born appreciating whisky; it is an acquired taste. The evolution of appreciating whisky starts with a dislike. At first, whisky will mostly taste the same unless you try them side by side, but over time you will begin to discern different flavours, you'll start to form a preference and begin appreciating whiskies that are far more complex. Having come full circle, that first whisky will no longer taste like every other whisky, as your palate has evolved through a form of education. This is the case with most things in life. If all we've ever known is fast food and ultra-processed foods, it will be nigh on impossible for us to choose an apple over an apple pie from McDonald's.

The Next Generations

My own children provide me with an abundance of hope. Our family is in a rare position given the work I do; this I am keenly aware of. But seeing their evolution as consumers from a food perspective has been reaffirming, to say the least. They experience everything the food system has to offer, as we do not deny them the joys of chocolate and ice creams. Sometimes they have Nesquik for breakfast, at parties they enjoy cupcakes with inches of artificially flavoured icing, and they both adore sweets – I indulge their adoration, bringing back country-specific varieties from my travels.

Of course, they also experience real flavour, diversity and mostly home-cooked food. Much of this has less to do with income and more to do with culture and tradition. Cooking at home is usually cheaper than eating out or ordering in. Diversity is a function of gastronomic knowledge – our ability to prepare and feel comfortable using a variety of vegetables and ingredients. Pulses, nuts, grains, spices, preserves; these are all the building blocks of a dish but also signs of a developed palate. I see it in Clem and Max daily: a reaffirmation that exposure at an early age does all the work for you. And the beauty is that it will never go away. No one can ever take away what you've learned; this is as true for academic knowledge as it is for your ability to discern between two glasses of milk.

Clementina complains regularly about the food at school – she is extremely aware of the processed foods they serve and is intimately aware of the seasons because we've been lucky enough to stick to seasonal produce at home. The other day, I cooked some white cabbage and Maximo told me it did not have an aftertaste. This is an eight-year-old telling you that this flavour is not complex and lacks length – he can recognise a lack of flavour through his exposure to good ingredients. If we can reintroduce diversity into the diets of children – and also adults – then our palates have been engineered over thousands of years to respond to that variety. If children can choose a hearty stew over a bar of chocolate when they're genuinely hungry, then we can be hopeful.

I see many similarities in my own exposure to food during my childhood. My memories are tied to healthy home-cooked food as well as the industrial hard-shelled tacos and the pink Big Mac sauce.

The point is that small changes are what we're after. Little by little, we can broaden our choices, get comfortable with where compromise is worthwhile as we navigate a difficult landscape that is made to sell us unhealthy food. I don't feel bad every time my kids have a Coke and neither should anyone else, but we must be aware of the intentionally addictive nature of these foods. Incremental shifts, small introductions of diversity, are all steps towards healthier eating. Together, those small steps we all take accumulate to become meaningful collective action.

Logic permeates nature and its systems. This is why flavour is a synonym for nutrition; there is obvious good sense in evolving to like things that make us feel good. This sense of well-being is so critical in explaining why we see children (and adults) reacting in the way Clem and Max did. Beyond the joy flavour brings, a diet founded on a diversity of unprocessed ingredients makes us feel good because it is good for us. While we enjoy that bag of Doritos and a hot dog at the stadium, too much of that food and our bodies react. We can fool our palate, but we cannot fool our bodies. High-energy, processed foods are severely lacking in minerals, vitamins, fibre and protein, all very important to our body's health. There is ample evidence that links diets high in processed foods with obesity, diabetes, high blood pressure and heart disease, and in children they have been linked with lower academic performance along with other troubling side effects like depression.

Food is there to nurture us. It is the fuel we need to stay alive but also the diverse and complex instrument that makes us feel great and allows our bodies to function at their optimum. We all want to feel well, there is no doubt about that. No one enjoys having an upset stomach or feeling rough and nauseated after a meal, but we have all felt this way sometimes. As we get older, the connection between what we eat and how we feel grows, it becomes more acute and we become more aware of it. No longer capable of handling alcohol in the same way we did in our twenties, we also have a tougher time handling unhealthy foods. This interconnectedness between flavour, nutrition and our well-being is designed into the human body. It is nature's way of telling us we have made the right choices when it comes to food. We need to lean in to these hardwired systems to leverage our instincts.

How do we recoup this lost knowledge so we can have greater ownership over our food choices? I'm not saying that demand on its own is the answer; alongside shifts in demand we need governments to change their priorities and focus on health, nutrition and sustainable ecosystems. We need the educational system to make the necessary changes to curriculums and school meals – the latter represents an often-overlooked educational moment. What our children eat five days a week is instrumental in developing informed and conscious consumers capable of making the right food choices. What we feed them in school is nothing less than a gastronomic education, and a decision to serve ultra-processed foods, or out-of-season fruits and vegetables, only serves to perpetuate the problem long into the future. The way that we stimulate and steer demand is by rebuilding our food knowledge, and education is priceless in doing so.

Georgina Webber was the deputy head teacher at Greenside Primary, a state school in west London, for over six years. One day, she reached out to me to express her admiration for Natoora's work and to share her own work at Greenside, which frankly I found to be even more noble. Over the years, her interest in farming and food systems grew to the point where she was convinced that she needed to introduce into the curriculum the concepts of sustainability, biodiversity, seasonality, soil health and even the cultural importance of food in our lives.

How these subjects were taught to primary students had the power of shaping the minds of tomorrow's consumers. She was shocked at how absent these crucial topics were from the curriculum. Georgina began farming in the courtyard so that students could build a connection to soil, plants and how we grow our food. She then seized school meals, wrenching them from a corporate third party and taking full control, ensuring students ate a seasonal, well-balanced diet, full of diversity and also plant-focused to serve all religious and dietary needs. This move away from animal protein not only helped eliminate dietary differences among students so they could all eat what was cooked irrespective of religion or culture, but importantly, it greatly reduced the cost of meals. After our initial contact, Greenside began sourcing their produce from Natoora, and we ensured we supported them by allowing them access to the best seasonal produce. We even helped them buy a bread oven so their chefs could bake fresh bread with well-sourced wholegrain flour.

Georgina did not stop at the kitchen. She hacked the curriculum, introducing this critical knowledge across most classes. In maths class they learned how to measure soil health, while in history they learned about the food cultures of faraway places. One day on a visit to Greenside, I was brought to tears when she opened the door to a classroom full of kids, barely seven years old, all chanting, 'Save our soils!' Can you imagine the impact these future consumers and leaders of tomorrow can have when they grow up with a deep knowledge of our food system, with a palate capable of making the right choices and a fervent belief in our need to preserve our natural habitats?

Education: it is the most effective way of stimulating the right choices. It needs to happen at all levels of society, yet younger generations are fundamental to this strategy – if we can raise better educated, more aware citizens, it paves the way for radical change. If our children don't want strawberries in the winter we will eradicate demand for them and, once economically unviable, we will no longer see them available year-round. Gastronomy must form part of the school curriculum, it needs to be reintroduced, particularly at the early stages of education. French lawyer and politician Jean Anthelme Brillat-Savarin, who wrote the seminal book *The Physiology of Taste* in 1825, said it most eloquently: 'Gastronomy is the knowledge and understanding of all that relates to man as he eats. Its purpose is to ensure the conservation of men, using the best food possible.'[2]

CONCLUSION

Flavour Can Spark a Revolution

The issues facing our food system are numerous and, given the consequent impact on natural ecosystems like the pollution of land and water, and even our health, they are some of the most pressing issues of our time. We have moved beyond the point of early thought and questions. It has become widely accepted that our current eating habits, the daily food choices we make, need radical change due to their outsized contribution to some of the planet's biggest problems. What is also evident is the complexity of the food system, a reality that must be considered when exploring both the challenges and, importantly, the solutions to revolutionising the food system.

Greenhouse gas emissions, soil degradation, degraded and erratic water cycles, the global obesity epidemic; these are but a few of the serious side effects of an industrial food system that prioritises cheap, nutritionally poor, water-rich food that is ultra processed with engineered flavour. The raw ingredients used in this scale-obsessed system are produced with complete disregard for these externalities. Such is the disconnect that they turn a blind eye to the degradation of their most precious production resource: soil. A sordid form of self-sabotage.

These years of abject disregard for the damage inflicted on our natural resources must end. The complicity across the food system among governments, supermarkets, seed companies and food manufacturers is punishing us and our children. Our post-war yield obsession, mixed in with governments' subsidies to sustain the chemical industry while weaponising commodities like corn and soy, turned natural foods like meat and vegetables into factory outputs. An industrialisation with wide ramifications for natural ecosystems through its polluting practices, alongside the persistent decline in the nutritional density of our food. Farms are larger, less secure and not valued. Neither are farmers or the incredible products they grow.

Confronting Complexity

All businesses that intersect with agriculture and our food system have a massive responsibility for the safeguarding of our future. Whether intentionally or not, they are inextricably bound up in multiple facets of our environment and economy. It is no coincidence that a system with such wide-reaching ramifications happens to be highly complex. The two are correlated and make meaningful systemic change all the more difficult.

The vastness of the system, with its multitude of perspectives and culturally ingrained habits, means there is no silver bullet. But this does not mean that change is unattainable; on the contrary, through consistent, collective changes to our consumption habits until we reach a tipping point, positive change is well within our reach. Change needs to come from all angles and actors in the food system, from consumers and governments to schools and universities, along with start-ups and the incumbent food producers. At every step of the chain, thousands of small changes must add up in order to reach this tipping point. The key with a demand-side shock is that it must flow from the end of the chain, from consumption, backwards into the supply chain all the way back to the farmer and ultimately the seed. Change must come from us, the consumer.

Flavour is our strongest ally in affecting change through demand. I believe I have built a strong case for that in these pages. I have seen first-hand its ability to win over consumers as well as its reliability in identifying production practices worthy of being labelled sustainable. Whether a peach, a chicken or a yoghurt, fantastic flavour can only be achieved when integrity is front of mind. We need to harness its power as a simple tool that guides our daily choices. Industrialisation has usurped our lives, rendering it flavourless. Recapturing the pleasure of eating something naturally delicious is worth some of the restraint needed to avoid out-of-season blueberries.

There is another dichotomy at play that must be broken. Told that dairy is unhealthy, we now opt for highly processed alternative milks. Plant-based diets are the way forward, yet we are offered ultra-processed plant-based foods that are just as noxious and irresponsible. Smoothies are the way forward, but little thought is placed on the impact of wanting blueberries in them year-round. What we are asking of the food system should be an integral, intuitive part of the equation, but this is difficult in a modern landscape where choice abounds and everything seems to have a catch. Again, this is where flavour comes in as a rule of thumb. What we should be asking for is flavour. In the winter months, flavour will lead us to oranges and grapefruits in our smoothies and granola rather than blueberries, whether we have a strong handle on seasonality or not.

There are few, if any, incentives to produce the quality of food we should be eating. The economic balance is tipped towards mass production and therefore our focus needs to be on shifting our demand. Without incentives like an innate desire such as Magnus's at the orchard, or government intervention, the only way to force the industrial machine into submission is by saying no. No to absurdly cheap food, to strawberries in the winter, to excessive junk food and consuming protein from concentrated animal feeding operations (CAFOs) on land and at sea. We should be voting with our wallets every time we eat or buy food, and we should be guided by our palate. By following flavour, we can trace a path back towards seasonality, improved nutrition, less pollution and respect for the environment.

A Flavour-first Food System

The unique experience I have gained in two decades of building capillary supply chains that reach deep into different regions around the world is at the heart of the bright future I see ahead of us. My experience, I believe, is invaluable, as it has grounded my ideas in the realities of what it takes to create financially sustainable sourcing networks in parallel with existing industrialised ones. Unless we are capable of moving product from farm to consumer efficiently, at its peak of ripeness, there is no demand that matters. If we cannot engineer a supply chain that grows, transports and sells flavour-first produce efficiently, we won't fix the system. The learnings from nations whose cultures foster quality must be transported to those countries where industrialisation has been more rampant.

The thoughts I have laid out in this book have taken years to form and coalesce into what I believe is a viable path forward. As complex as the system is, and as multifaceted as the solutions need to be, the top of the funnel – demand, to be precise – can be addressed through beautiful simplicity. It needs to be simple to cut through the marketing, confusing labels and contradictory advice. A simple tool is all we need to make the right choices. Flavour is that tool, and shifting that demand is the answer. A complete reformulation of how demand funnels through the supply chain.

What does a flavour-first food system look like? Our food system can be viewed as a collection of supply chains. Thousands of organisations around the world have built ways of moving product from farms into restaurants, retail stores and food manufacturing facilities, and they range from hyper-local to global. There is inbuilt infrastructure – roads, trucks, ships, warehouses, wholesale markets – that enables food to move from farm to consumer in a multitude of ways. It is imperative to harness the existing infrastructure; there is no other way. The cost of moving product outside of these networks is too high and inefficient.

What needs to change is the quality of product that travels through the global infrastructure that moves product around. By utilising the same delivery mechanism, just like changing a keg of beer, you can switch from industrial lager to beautiful craft ale effortlessly. Changing the product requires alignment across the entire network of supply chains from top to bottom, one supply chain at a time.

In this transition phase, a new network of food producers, a new supply chain, must be built to stimulate flavour-led demand. Getting the transition underway requires organisations that either rebuild supply chains or graft new ones onto the existing infrastructure to allow this demand to flourish and grow. It is an interesting paradox; without access to flavourful oranges, it is pretty damn hard to demand them, so, initially, these new supply chains must undertake the work of creating access to the right food. By providing access you open the possibility for customer experience, as tasting quality is fundamental while education supports and accelerates the transition.

I can't stress this enough: finding the product is half the challenge, the other is developing a customer base that wants better. Traditionally, we don't think of the consumer as being part of a supply chain, but rather as the destination of the product that moves within it. If we are to transition to a flavour-first food system, we need to see the end user as an active participant in the chain.

The interconnectedness of all actors in a specific supply chain is therefore vital to acknowledge, and individuals or organisations building them must align farmers, themselves and their customers. It is only when all the actors are aligned that ripe fruit can move through from orchard to retail shelf, or for the right value to be apportioned to an amazing carrot and for the right farming practices to be employed.

As you build a wider community of flavour enthusiasts, you can find more farmers to work with. The larger the community, the more buying power and the easier it is to move product through the chain. It is an iterative process that gains momentum and builds upon itself. This is why chefs play a pivotal role in building the food system of the future – they are the early adopters who act as the ignition spark and have the ability to move great volume, as well as inspire and influence everyday consumers.

At the other end, we need an extremely capillary farming ecosystem. Capillary because it allows the system to reach deep into regional pockets and pull from a vast, diverse network of farms of all sizes and practices. Many of these farms need to be accessible-scale operations that prioritise specialisation and scale, therefore producing better quality in volumes that open up existing delivery infrastructure – this provides access. Without accessibility we can't reach consumers en masse to enable flavour-driven demand to filter through into the food system. Central markets, and other nodes, are integral to the food system of tomorrow because the strength of these supply chains is their capillary nature, viable only through these nodes, which act like the largest veins (the body's vena cava), capable of consolidating all the product and transporting it to the heart. This model allows us to tap into the strength of rich cultural heritage, and importantly overcome the gulf present in many great agricultural areas that don't have sufficient local demand to sustain quality, as we see in Mexico and the work of Tamoa on heritage corn, beans and chillies.

As these new supply chains take shape and more product is easily accessible to a wider audience, they begin to funnel more demand into the right farming and production of our food. Supply chains coexist, there are arguably thousands of different supply chains that together represent the food system as a whole, but as momentum picks up, existing chains that move industrially farmed produce will react to the change in consumer desires and begin their own transition towards better flavour.

Inspiring Change

My own love of fruits and vegetables aside, there is good reason for dedicating my professional life to supporting the transition towards a flavour-first food system. I see a path forward, a brighter future.

I have learned to value products through the direct contact I have with the hundreds of amazing farmers we work with. Appreciating their struggles and seeing the love they put into their growing is a big part of that. At Natoora, we aspire to engage consumers with the magic and beauty of a perfect plum or the sexiest mustard greens. We have created ways of taking produce out of context as a way of showcasing it, and we dedicate a lot of time to communicating the physical characteristics that people should look out for when seeking quality.

Irrespective of where people buy their food, we are conscious of our responsibility as well as our ability to spread our knowledge for the common good. The stories I have shared here are an evocative way to create that same sense of reverence for a humble plum. I know that as we all learn to revalue simple ingredients we will be more willing to try them and, where we can, be prepared to pay a higher price for them. When you see the beauty and artistry in fruits and vegetables, I guarantee you will revalue them. Look at a pear with renewed interest, take some extra time to appreciate its character traits, let it speak to you through its physical appearance even before you take a bite.

We need to be more prescriptive with seasonality. As challenging as this may seem, even supermarkets offer seasonal produce alongside unseasonal items. Following the basic tenets of seasonality is a great way to start – local asparagus in the spring, berries in the spring and summer, stone fruit in the summer, apples and pears in the autumn, and citrus in winter. Vegetables are a little harder to decipher but chances are that local sweet peas still in their pods will be seasonal, whereas off season they will come from abroad. And of course, let flavour guide you. That is ultimately the lesson. Actively pursuing a more seasonal diet will re-engage you with nature and its rhythms. Flavour will go up, you'll enjoy food more, you'll feel better and importantly you will reconnect your body and its well-being with natural cycles. No longer will you crave peaches in winter. Your cooking will flourish.

We must move away from processed and ultra-processed foods. We don't need to eradicate these foods from our lives, but we must reduce them if we are to break away from the industrialisation of nutrition. If our children recognise and understand their addictive nature and harmful effects, they will use good judgement and be responsible consumers of these products. We must be aware that this is not food; it is not there to nourish or sustain us. And we can only do this by using flavour to guide our food choices and demand different. There is no other way.

While I am totally in favour of sustainably reared protein and believe it is an integral component of a healthy farming ecosystem, shifting our diets towards more plant-based foods has the greatest impact in reducing our own carbon footprint. This is because plant-based food uses less resources than animal protein. For one, we consume too much meat because it has been commoditised. Subsidies, industrialisation and the quest for scale have created an industry that rather than producing healthy, nutritious animals that support soil health and foster intelligent natural farming systems, sees animals as factory outputs. How can we continue to produce meat as cheaply as possible with no regard for its nutritional value or the grave consequences of the process to our land and natural habitats? We should all consume less meat, and the meat we do eat should be higher welfare. Plants need to make up a higher proportion of our caloric intake to rebalance the food system and steer us away from these inhumane protein factories. Be careful, as not all plant-based diets are the same. If you eat mostly plants but they come in cartons and boxes, all you've done is shift consumption from highly processed food to highly processed, animal-protein-free food. Let me be clear: this is not a positive outcome. Is it better? Yes, but not nearly good enough. Not good enough for your health, and not good enough for the planet. Flavour is what helps us navigate the complex web of choices.

Seek out flavour, follow seasonality, move away from processed foods, lean in to unprocessed plant-based ingredients, and help spread the word. In other words, be complicit in educating ourselves for the revolution depends on it.

Hope

A blind belief that a monumental shift is possible has given me the strength to endure twenty years of fighting for radical change in our food system. Twenty years obsessed with flavour carries me through the difficult moments with unwavering commitment. The joy of flavour and the importance of the cause are all I need to continue down this path full of hope. A revolution of our food system is at once imperative and eminently possible.

Flavour has been capable of burning memories into my core along the voyage that I've taken with Natoora. The first time I smelled and tasted bergamot. Eating a green mandarin off the tree in Sicily for the first time in my life. The first meal I had at I Rizzari, and the countless ones after. The day we found Domenico and his phenomenal white peaches. That night at Sean's house when we spent the whole evening talking about the future on the day we met. The day I first set eyes on Emerenziana and her husband Antonello's works of art. It is very powerful.

If we want to see the dawn of a new food system, we must all contribute towards it. We all have an active role to play in shaping the food system of tomorrow. Our choices at breakfast, lunch and dinner are small revolutionary acts that, when taken collectively across a community or society, have tremendous power. Next time you're at a train station eating a tasteless fruit salad, let your mind wander. Recognise that the melon is out of season, tasteless, has no nutrition and the soil it grew on is likely worse off. You should ask yourself, should I be eating this? Sometimes there is no option, but at other times there is. And even if in that moment you have no alternative, your awareness will inform your future choices.

Flavour is the best compass we have, the simplest to use and the most enjoyable. We must reintroduce flavour to the centre of the food system. We must return power to the consumer. And we must break the industrial food cycle. A flavour-driven revolution that will deliver the flavour-first food system we so desperately need.

This is the moment when I hand over to you.

Acknowledgements

While this book is the culmination of two decades of work, it would not have seen the light of day without the support of many incredible individuals.

My wife, Daisy, for always believing in me, in my vision for a better world and for her unwavering support, which has allowed me to devote myself to Natoora. Perhaps more importantly, she has made the last seventeen years together an incredibly fun ride. This book would not have been possible without her.

My children, Clementina and Maximo, for the daily joy they bring to our family, for their interest in my work and for believing in the power of seasonality. I trust one day they will find their own calling.

My brother, Alejandro, without whom I would not be where I am today. He supported me in the early years and has a generosity beyond words. He's an incredible brother whom I will forever be indebted to and who happens to make the best ice cream in the world!

My mother, who sacrificed so much to make me who I am. She dedicated her life to me and my brother, and our families, selflessly. No words could ever do justice to or repay her unwavering commitment, and for how hard she fought for us. Any good I can accomplish through this book belongs to her as well, as it reflects her efforts in giving me the tools and confidence to be free. She's one of a kind.

Alfredo Castelli for all the music, but more importantly, for being a strong support to our family and being by my mother's side for all these years.

Alberto Medus who was a father to my brother and me, and through his spirit and character taught us the value of family and moral fortitude. He left this earth too soon yet his influence was enormous and will endure forever.

My father, for sharing his cooking and cooking with me, and in doing so instilling in me a special connection to food and its preparation. During some of Natoora's most challenging times, the two of us had many valuable conversations.

Federico Cervellin, as Natoora would not be the amazing organisation it is without his influence and the farming relationships he has nurtured. He has made me a better leader, has been the source of some of the most visionary ideas, and his blood, sweat and tears are forever etched into Natoora's culture.

Mary Coote for her tremendous commitment and dedication. This book would not have been published without her and her belief in its importance. She helped me find my voice through the many conversations and edits over the years, crafted all the book proposals, and is responsible for defining the structure in this book. Our conversations during its writing and her endless positivity kept me going to the end.

Everyone at Natoora, past and present, for being part of one of the few for-profit organisations in the world that is truly fighting for a more positive future. Without them none of this would be possible. There are too many to name, but I am honoured to work with each and every one.

There are a handful of colleagues who were there in the early years, and whose immeasurable efforts carried us through the most difficult moments, ensuring Natoora stood the test of time: Tim Ballard, Katherine Miller, Rafal Galant,

Luis Marques, Sabrina Zanellato, Sergio Pinho Vaz Pinto, Robert Watorek, Joel Santos, Nelio Ferreira Fernandes, Bizu Teshome, Carlos Eduardo Rodriguez, Jose Santos, Antonio Jesus and Paolo Coquinhas. And a handful no longer working with us: Pedro Mota, Ignacio Pestana and Andre Dos Santos.

I want to thank Ruth Rogers and the late Rose Gray for believing in me so early on, and for being wonderful supporters and friends. As well as Joseph Trivelli and Sian Wyn Owen and the entire River Cafe family. So much of this started with them.

To Ed Wilson, Antonin Bonnet, Theo Randall, James Lowe, Brett Graham, Marco Torri, Isaac McHale, Claire P tak, Claude Bosi, Tom Adams, Andrew Clarke, Stevie Parle, Sky Gyngell, Graziano Bonacina, Gianluca 'Gianni' Canna, Jeremy Lee, Andre Garrett, Jacob Kennedy, Francesco Mazzei, Michele Lombardi and all the other incredible and inspiring chefs whom I've had the fortune to work with.

Muna Reyal for believing in the book since our first meeting, as well as being a kind and very effective editor. The book is far better thanks to her.

Rose Baldwin and the team at Chelsea Green for giving me this opportunity, seeing the need for this book, and giving me the freedom to stick to my original vision.

My dear friend Nicolas and his amazing wife, Lea, who opened their home in Montmartre during the early days of Natoora, when I was regularly travelling back and forth to Paris. Their generosity, and very comfortable living room sofa, contributed to Natoora surviving those frugal days. That period will forever be part of the Natoora history.

Sean O'Neill for being an inspiration and boundless source of hope, opening more layers to life, and having an eye for the tastiest mushrooms.

Daniele and Claudia for hosting us every time we go to Sicily, opening their home and welcoming us with incredible food and, of course, for producing some of the world's most delicious olive oil.

I need to thank some farmers and other individuals for believing in Natoora when we were just getting started because without their belief, without their help, we would not be here today: Francesco Tallarita, Alessandro Pirola and family, the Daniotti family, Oli Baker, Bruno Parisi, Ferdinando Vinaccia, Marc Semenzato, François Blanchon, Domenico Magliozzi, Michele Todisco, Pietro Biffi, Thierry Donzelot and Gwen Martin.

Finally, all the farmers and producers who have made the last twenty years a dream. The produce, the meals, the shared belief and commitment to flavour have demonstrated that a better future is within our reach.

Notes

Introduction

1. Hattie Burt and senior policy & international projects officer at Action on Sugar, 'Why Supermarkets Are "Leanwashing" on Sugar Reduction', *Grocer*, 11 April 2023, https://www.thegrocer.co.uk/health/why-retailers-are-leanwashing-on-sugar-reduction/678060.article.

2. 'Food and Beverage: Supermarkets Remain King in Food Retail', Statista, accessed 04 January 2024, https://www.statista.com/markets/423/topic/451/food-beverage/#statistic2.

3. Anne-Marie Mayer, 'Historical Changes in the Mineral Content of Fruits and Vegetables', *British Food Journal* 99, no. 6 (1997): 207–211, https://doi.org/10.1108/00070709710181540.

4. Donald R. Davis, Melvin D. Epp and Hugh D. Riordan, 'Changes in USDA Food Composition Data for 43 Garden Crops, 1950 to 1999', *Journal of the American College of Nutrition* 23, no. 6 (2004): 669–682, https://doi.org/10.1080/07315724.2004.10719409; Alex Jack, *America's Vanishing Nutrients: Decline in Fruit and Vegetable Quality Poses Serious Health and Environmental Risks* (Becket, MA: Amberwaves, 2005), https://www.betterbones.com/wp-content/uploads/2016/03/Americas-vanishing-nutrients-Decline-in-fruit-and-vegetable-quality-poses-serious-health-and-environmental-risks.pdf.

5. Mark Schatzker, *The Dorito Effect* (New York: Simon and Schuster, 2016), 29.

6. Elizabeth K. Dunford, Donna R. Miles and Barry Popkin, 'Food Additives in Ultra-Processed Packaged Foods: An Examination of US Household Grocery Store Purchases', *Journal of the Academy of Nutrition and Dietetics* 123, no. 6 (2023): 889-901,

https://doi.org/10.1016/j.jand.2022.11.007; Sarah Boseley, '"Ultra-Processed" Products Now Half of All UK Family Food Purchases', *Guardian*, 2 February 2018, https://www.theguardian.com/science/2018/feb/02/ultra-processed-products-now-half-of-all-uk-family-food-purchases.

7. Zoé Colombet et al., 'OP12 Social Inequalities in Ultra-Processed Food Intakes in the United Kingdom: A Time Trend Analysis (2008–2018)', *Journal of Epidemiology and Community Health* 76, supplement 1 (2022): A6-A7, https://doi.org/10.1136/jech-2022-SSMabstracts.12.

8. Filippa Juul et al., 'Ultra-Processed Food Consumption among US Adults from 2001 to 2018', *The American Journal of Clinical Nutrition* 115, no.1 (2022): 211–221, https://doi.org/10.1093/ajcn/nqab305; Lu Wang et al., 'Trends in Consumption of Ultraprocessed Foods Among US Youths Aged 2–19 Years, 1999–2018', *JAMA* 326, no. 6 (2021): 519–530, https://doi.org/10.1001/jama.2021.10238.

9. Schatzker, *The Dorito Effect*, 17.

Chapter One

1. Graeme Paton, 'British Public "Ignorant about Seasonal Fruit and Veg"', *Telegraph*, 10 August 2014, https://www.telegraph.co.uk/news/earth/agriculture/food/11024484/British-public-ignorant-about-seasonal-fruit-and-veg.html.

Chapter Four

1. David R. Montgomery and Anne Biklé, *What Your Food Ate: How to Heal Our Land and Reclaim Our Health* (London: W.W. Norton & Company, 2022), 24–25.

Chapter Five

1. H. Garrison Wilkes, quoted in Charles C. Mann, *1491: New Revelations of the Americas before Columbus* (New York: Vintage, 2006), 221.

2. John S. Allen, 'Food and Memory', *Harvard University Press Blog*, 18 May 2012, https://harvardpress.typepad.com/hup_publicity/2012/05/food-and-memory-john-allen.html.

3. Cecilia Nowell, 'Is Eating Local Produce Actually Better for the Planet?', *Guardian*, 7 June 2023, https://www.theguardian.com/environment/2023/jun/07/is-eating-local-better-environment.

4. *Oxford Advanced American Dictionary*, s.v. 'craft', 6 March 2014, https://www.oxfordlearnersdictionaries.com/us/definition/american_english/craft_1.

5. Michaela Clarissa Theurl, Helmut Haberl, Karl-Heinz Erb and Thomas Lindenthal, 'Contrasted Greenhouse Gas Emissions from Local versus Long-Range Tomato Production', *Agronomy for Sustainable Development* 34, no. 3 (2014): 593-602, https://doi.org/10.1007/s13593-013-0171-8.

Chapter Eight

1. Howard Elitzak, 'Food Spending in Relation to Income' in *Food Cost Review, 1950-97*, Agricultural Economic Report No. (AER-780), (Washington, DC, US Department of Agriculture, 1999), 20, https://www.ers.usda.gov/webdocs/publications/41035/15331_aer780e_1_.pdf?v=0.

2. Eliana Zeballos and Wilson Sinclair, 'Average Share of Income Spent on Food in the United States Remained Relatively Steady From 2000 to 2019', *Amber Waves*, 2 November 2020, https://www.ers.usda.gov/amber-waves/2020/november/average-share-of-income-spent-on-food-in-the-united-states-remained-relatively-steady-from-2000-to-2019.

3. Richard J. Johnson, Laura G. Sánchez-Lozada, Peter Andrews and Miguel A. Lanaspa, 'Perspective: A Historical and Scientific Perspective of Sugar and Its Relation with Obesity and Diabetes', *Advances in Nutrition* 8, no. 3 (2017): 412–422, https://doi.org/10.3945/an.116.014654.

4. Johnson et al., 'Perspective', 412–422; Judith Hallfrisch, 'Metabolic Effects of Dietary Fructose', *The FASEB Journal* 4, no. 9 (1990): 2652–2660, https://doi.org/10.1096/fasebj.4.9.2189777.

5. 'How Much Sugar Is Too Much?', American Heart Association, 22 August 2016, https://www.heart.org/en/news/2023/05/23/kids-and-added-sugars-how-much-is-too-much.

Chapter Nine

1. Devi B. Dillard-Wright, 'Technology Designed for Addiction: What are the dangers of digital feedback loops?', Psychology Today, 4 January 2018, https://www.psychologytoday.com/us/blog/boundless/201801/technology-designed-addiction.

2. Ole G. Mouritsen, 'The Emerging Science of Gastrophysics and Its Application to the Algal Cuisine', *Flavour* 1, no. 6 (2012), https://doi.org/10.1186/2044-7248-1-6.

Index

Index

Index

About the Author

Franco Fubini is at the forefront of a fundamental revolution in the way we all eat. As founder and CEO of Natoora, the leading seasonal food distributor dedicated to fixing the food system by connecting over five hundred independent farms with the world's best restaurant chefs, Franco has shaped the food industry in London, Paris, New York, Miami, LA, Copenhagen, Malmö and Melbourne. His unique approach to seasonality and commitment to direct sourcing has been raising the bar on flavour for the past twenty years.

Driven by his belief that by engaging people with the real flavour of fruit and vegetables, we can collectively transform the way that food is being farmed and supplied, Franco has built a resilient supply chain founded on flavour, transparency and direct relationships. A professor of sustainability management at Columbia University in New York, Franco is also a jury member for the Food Planet Prize, the largest environmental prize in the world focused on food systems solutions.